WITHDRAWN

"An audacious and exhilarating book."
—SYDNEY MORNING HERALD

"A whirl of a book. . . . Any page will get you hooked."
—NEW SCIENTIST

"A brave and novel rage against the machine of time, this book oozes ideas as rich as a literary death-by-chocolate. Savor a spoonful at a time."
—GIBBY ZOBEL, BIG ISSUE

"Like the seminal socialist, feminist and ecological works, A Sideways Look at Time articulates what thousands have felt but no one has been able to put into words. Suddenly, shapeless concerns are brought into focus. Outrage takes the place of confusion, fascination displaces complacency. Cheeky, intelligent, always gripping, this book reintroduces us to a dimension we've utterly neglected. It will be the opening salvo in a new battle over the human spirit."
—GEORGE MONBIOT, AUTHOR OF CAPTIVE STATE

"Ambitious . . . playful, feminine, spontaneous and hedonistic."
—THE ECONOMIST

"Both revolutionary and a real pleasure to read. Certainly this must become one of the principal texts of anyone campaigning for social or environmental justice."
—THE ECOLOGIST

"A wonderful book, full of illuminating juxtapositions. Jay Griffiths has invented a new literary genre."
—RICHARD GOTT, AUTHOR OF IN THE SHADOW OF THE LIBERATOR

"Splendid, extraordinarily wide-ranging, impassioned and often outrageously witty."
—NEW INTERNATIONALIST

"An extraordinary book that challenges us to take a very different view of time."
—GREEN FUTURES

"A delight. This gripping book, inspired by one too-fast trip down the motorway, manages to take a holistic overview of time, rattle clockwatches and praise children for their natural anarchic ability to maintain tradition. All this done with an energetic panache that will leave you mentally richer."
—EARTHMATTERS

A SIDEWAYS LOOK AT TIME

JEREMY P. TARCHER · PUTNAM

a member of Penguin Putnam Inc.

New York

JAY GRIFFITHS · A SIDEWAYS LOOK AT TIME

Lyrics for "Designer Kidz" reprinted with permission of Theo Simon.

Quotation from *Poems* by Dafydd ap Gwilym, edited by Richard Morgan Loomis. Reprinted with permission of The Center for Medieval and Early Renaissance Studies.

Most Tarcher/Putnam books are available at special quantity discounts for bulk purchases for sales promotions, premiums, fund-raising, and educational needs. Special books or book excerpts also can be created to fit specific needs. For details, write Putnam Special Markets, 375 Hudson Street, New York, NY 10014.

Jeremy P. Tarcher/Putnam
a member of
Penguin Putnam Inc.
375 Hudson Street
New York, NY 10014
www.penguinputnam.com

Copyright © 1999 by Jay Griffiths
Previously published in Britain as *PIP PIP* by HarperCollins Publishers. First American edition published by Jeremy P. Tarcher, 2002.

Library of Congress Cataloging-in-Publication Data

Griffiths, Jay.
[Pip pip]
A sideways look at time / by Jay Griffiths.
p.   cm.
Originally published under the title: Pip pip.
ISBN 1-58542-184-7
1. Time measurements—Popular works.   I. Title.
QB213.G75   2002   2002025369
529'.7—dc21

Printed in the United States of America
10   9   8   7   6   5   4   3   2   1

This book is printed on acid-free paper. ∞

Book design by Gretchen Achilles

CONTENTS

If, on the one hand, any of the following apply to you:

If you suspect there's more to time than clocks . . .

If you do *not* think that time is money . . .

If you would have laughed overhearing a child say "I'll do it in five minutes . . . Is that today?" . . .

If you have a sneaky feeling that other cultures might have wise, rich and elegant ideas of time . . .

If you have ever wondered whether linear and cyclical time could conceivably have something to do with gender . . .

If you have ever felt that Western modernity's time is coercive, crushing and overwound . . .

If you think an analysis of time could, or indeed should, include art, advertising, philosophy, literature, anthropology, history, sociology, music and myth . . .

        . . . Then please read on.

If, on the other hand, the following apply:

If you think a book about time can only mean a history of clocks . . .

If you want an exhaustive, chronological reference book . . .

If you think that time is money . . .

If you'd rather count time than cherish it . . .

If you like clocking-in . . .

If you like everyone else clocking-in . . .

If you simply can't bear childhood, carnivals, mirth, sex, nature, dawn choruses and May Day . . .

        . . . Then let me politely head you off at the pass.

·   ·   ·

There are many excellent books on time. Historians writing on time include E.P. Thompson, Anthony F. Aveni, David Landes and Carlo M. Cipolla. G.J. Whitrow and David Ewing Duncan have written interesting books on the history of calendars. In scientific literature, you have Stephen Hawking, Stephen Jay Gould, Peter Coveney and Roger Highfield, Ilya Prigogine and others. Among sociologists, Barbara Adam is superb.

So, why this one? It's a book about how people experience time, not an abstract academic study. It's an argument, not a reference book. It began in a euphoric moment, thinking about speed. Speed fantastic but fatal, speed exhilarating but frightening, fabulous but fascistic.

It continued with an instinct that "time" was a highly political subject, and that modern Western time is a subtle but profound example of cultural imperialism.

It proceeded with a warm appreciation for the ways people all over the world have pictured time, and for time's rootedness in nature. It travelled with me, to India, Thailand, America, Canada and New Zealand.

It involved reading hundreds of books and articles in the British Library, all of which were interesting, stimulating and useful, on very scattered aspects of the subject. None of them, though, sought to put all this information together, to look at time from a fully cultural perspective. Furthermore, though there were glittering gems of fury about certain subjects (Paul Virilio on speed, for example, and various critics of progress), no writer seemed to have made a single, cohesive polemic as I wished to do, a broadside against *all* the misuses of time in modern Westernized societies. And a manifesto for time to be seen as something more extraordinary, strange, sensual—even erotic—than our dominant definitions allow.

In such a book, using the dating devices B.C. and A.D. ties time into an inappropriately rigid Christian context, so I have instead used B.C.E. (Before Common Era) and C.E. (Common Era) respectively. The chapters are arranged thematically. The first looks at the present, too often enclosed or tamed by clocks, and at the variety of ways that "the now" is portrayed across the world. Chapter two is about speed moving to one on the past, and one about carnivals and rhythms of time. Following that is a chapter on women's time, then one on gender, one on power and one on money. The next chapter discusses progress, then there's one on the future, one on nature and one on death. The last chapter looks at "wild time," time set free, unenclosed, time as it could be conceived. This was partly written in one of the last remaining wildernesses on earth, a wild land and a wild river—the perfect setting for wild time. In a concession to linear time, the chapters are numbered sequentially, one to thirteen. The thirteenth chapter represents the thirteenth hour—the ultimate unclocked wild hour.

. . . . *pip pip* . . . .

Ah, now! That odd time—the oddest time of all times; the time it always is . . . by the time we've reached the "w" of "now" the "n" is ancient history.

—MICHAEL FRAYN

An artist, a woman and a gridded screen. The woman is naked and entirely sensual. She reclines away from the artist, smiling a very knowing smile, and—if he looked—the artist would see her left hand is teasingly, playfully, almost touching herself. What of the gridded screen? The artist has set it between himself and the nude, so he can accurately plot her measurements and proportions. So rigidly preoccupied is the artist with the grid, that on the paper before him is no woman at all, no knowing smile, no thigh and no moist finger, just the straight lines on the page, the frets of a grid. Albrecht Dürer (1471–1528) made an engraving of this, the "Artist Drawing a Nude through a Gridded Screen." Looking at it once, it occurred to me that this is how modernity sees time; that we are so preoccupied with our gridded, subdivided constructions of numbered measurements that we lose sight of the gorgeous, lifeful thing itself. Modernity knows the strut and the fret. But not the hour.

Time's measurement is everywhere. The gridded screen surrounds you like barbed wire. Leaving Washington, New York, London, Madrid, Berlin or Paris: at the airport, every transaction, each ticket and money exchange is timed. Around Heathrow, as at any other major airport, there are clocks on corporate buildings and hotels, blinking the date and time, down to tenths of seconds, compressing time. Closed-circuit television (CCTV) times your

1

progress along streets. Stop to buy a shoelace and your receipt will tell you the date and time—to the minute. The pips of time-measurement are spat out by the radio. (And if these are the pips, where is time's fruit?) Traffic lights run on timed schedules. Tele-phones tell you the time, whether or not you want them to (one Swiss telephone buzzes every twelve minutes, so you can measure your wordage in time); breakfast TV has a constant clock; there are pingers on cookers and eggs have sell-by dates printed on their very shells. Urban modernity lives under an assault of clocks. Alarm clocks put the frighteners on sleep: the first thought in so many people's minds, every single waking day is "What's the time? Am I late?" Digital clocks with digital seconds seem to speed time, relentlessly tightening deadlines. Increasingly-clocked schedules puncture time. In the U.S. paper was the traditional gift for a first wedding anniversary: now it is clocks. Crystal was the tradition for the fourteenth anniversary: now it is watches.

Cities more than anything create the clock-time of moder-nity, constructing a whole world based on the artifice of the clock and calendar. The heart of urban modernity is the clock and it is beating faster and faster. Strut, fret, strut, fret. People speak of the frenetic pace of modern life. Speed, as this book will show, has the whip-hand in today's society. Speed is deceptive and alluring, adrenaline-pounding and cruel, fantastic and fascistic; everything is speeded up, from relationships to ever more temporary jobs, from fast food to fast clothes and fast knowledge, foreshortening time. But as well as being speeded up, time is broken into ever smaller pieces. Diaries are printed with gridded time in fifteen-minutely appointments. A clock squats in the corner of many computer screens, while computers themselves split time and di-vide it into picoseconds and nanoseconds, far beyond any human sense of time, or human need, division for long division's sake.

Minutes (invented by the Babylonians) were little used until the Industrial Revolution, which "needed" more exact time-measurement. Thomas Hardy in *Tess of the D'Urbervilles* noted their advent thus: "Tess . . . started on her way up the dark and crooked lane or street not made for hasty progress; a street laid out before inches of land had value, and when one-handed clocks sufficiently subdivided the day."

Since then, time has been increasingly divided and subdivided, an obsession which is a feature of modernity, to divide time—that infinite and beautiful thing—into the mere infinitesimal decimal, as scientists today use femtoseconds, a millionth of a billionth of a second. Childbirth is timed mechanically. A child's ability to learn is measured not by depth of understanding but by speed of progress. Work is usually costed by hours. Quality time only thinks it's so special because dominant time is quantitative, counted and accounted. Everything is timed. It even happens in the fridge. Today, milk cartons are being produced with the usual Best Before Date. Best Before 10/3. (Well, if you must.) But then some of them add "at 09:52." So at 09:53 it will all of a sudden curdle itself to a yogurty oblivion.

Listen carefully.

9,192,631,770 oscillations of the cesium atom define a second. Count. There are 86,400 seconds in a day and every one of these is officially pipped off, day after day, by the time-measurement division of the National Institute of Standards and Technology in Boulder, Colorado, broadcasting *the* time.

Listen very carefully.

Since there are 86,400 seconds in a day, there are, therefore, 31,536,000 pips per year. Count. Count. With the right receiver you can convert the radio signals into the pips of seconds and count them. There is a wail at every second and a flurry around the minute.

Count, in fact, wherever you are, for *the* time is universal. This is *the* time, since on the first of January 1972 the definition of a second was designated as the atomic second and Co-ordinated

Universal Time (abbreviated to UTC or UT) was set by international agreement. Before that, the second was defined as 1/86,400 of the mean solar day, i.e., a rotation of the earth.

If you had listened properly, which you obviously didn't, on 06/30/97, or on 12/31/98, you would have heard 86,401 seconds, that extra one a leap-second. Roughly every year, a leap-second is added to realign *the* time with that of the earth—it is added to "accommodate" the earth's unreliable time, for there is no natural gearing between it and the atomic clock's pip pip pip. For the earth, you see, is too inaccurate for modernity's time measurement, because its spin changes by up to a thousandth of a second in some years. A thousandth of a second, indeed, tut tut, how unpunctual the earth. (It is slowed down by the gravitational pull of the moon. But the "atomic fountain" clock, the NIST F-1 at Boulder, "is entirely free from the timing errors created . . . by the earth's irregular movements in orbit.") So the timekeepers of today must tell *the* time from above, or outside the earth itself, insisting, with the caesium clock, with speaking clocks and the whole notion of GMT or UT, that there is one time, abstract, universal, mono-time: *the* time. There is no such thing.

Switch off the pips. Bury your alarm clock and digital watch. Forget the global calendar of the worldwide mono-day, 10/06. And come with me instead to visit the Karen, a hill-tribe in the forests of Northern Thailand, and see how such a forest can be teeming-full of times. Or to Indonesia on the night of the "little pig moon." Or to the Andaman forests and smell your way through the scent-calendar. See cow-time and bee-time, coconut clocks, Watermelon months and the month of the Snowblind. See how the seas and oceans are full of time, culturally and physically; time and tide, in rhyme and rime chiming in the mind of humanity. There are thousands of times, not one. To say any one time is *the* time is both untrue and highly political.

So let's go. Slam the fridge door on milk carton-time and see cow-time instead. Leave "the" time behind you and see a few of the myriad times across the world.

The Karen always know the time. Living with them for six months it became clear to me that the only person with a watch and the only person who could never tell the time was, well, myself. To the Karen, the forest over the course of a day supplied a symphony of time, provided you knew the score. The morning held simplicity in its damp air, unlike the evening's denser wet when steam and smoke thickened the air. Backlit by sun, a huge waxy banana leaf at noon became green-gold stained glass, cathedralizing time. Barely one of my hours later, it was just a matt, bottle-green leaf, useful verdure, a plate for rice, a food-wrapper. Birds sang differently at different hours and, while the soloists of life are always with us, the whole orchestra of the forest altered, shifting with the sun's day, all the noisy relations between birds, animals and insects, making chords of time played in all the instrumental interactions. Western time seems a thin, thin reedy peep of a thing by comparison.

The Karen always know where they are and when they are, how far they are from sunset or home: for time and distance are connected in the Karen language: *d'yi ba*—soon—means, literally, "not far away." Sunset, therefore, could be expressed as "three kilometers away," because the only way of traveling is to walk, which takes a known length of time.

There was a similar indivisibility between social time and action. I—eventually—learned not to ask "When will such and such happen?" When will a wedding, for instance, take place? For the no-answer-possible smiles replied. It happens when it happens: the doing of a thing and its timing are indivisible, the action is not jostled into the hour, but the hour becomes the action and the action becomes the hour. What to me was a distinction between the hour and the act was, to them, tautology, half artificial and half daft.

To the Karen, the whole forest was a clock. To the Kelantanese of Malaysia, the coconut can be used as a stopwatch to time a game, for, by drilling a hole through half the shell and putting it in water which then bubbles up into it, the coconut eventually sinks and time is up. Black Elk of the Oglala Sioux speaks of the June moon as the "Moon of Fatness," when all of nature is at its ripest. In parts of Indonesia, two of the waxing nights of the moon are called the "little pig moon" and the "big pig moon"; the nights the Western calendar would call the eleventh and twelfth nights of the moon's growing. They are so called because these are the nights when pigs and piglets get moonstruck, tickled pink by moonlight at night and squeakily overexcited, they bust out of their pens and go for a cavorting, giggly, piggy scramble in the fields. The waning nights, when the moon gets stubbier, are called the nights of the "long tree trunk" and "short stump."

To the Ugric Ostiak in northern Siberia one month is called the Naked-Tree month followed by the Pedestrian month (when people must journey not on horseback but on foot, and that gingerly, over ice). The month of Crows (also called Wind month) comes before the Spawning month, then follow the Pine-Sapwood month, the Birch-Sapwood month and the Salmon-Weir month. Across the world the nature of each moon-month is characterized and, through each people's names for the months, you can "see" the specific landscape they inhabit. (Through "12/87" or "6/03" what can you "see" except a global suburbia?)

The Natchez tribes of the lower Mississippi river valley have months including the Deer month and the Strawberries month, the Little-corn, Watermelon, Mulberry, Nut, and Great-corn months, the Turkey, Bison, Bear and Chestnut months. The calendar of the Malagasy people of Madagascar offers the Gourds-Flower month and the Bulls-Seek-Shade-of-Sakoa-Tree month while the next is the Guinea-Fowls-Sleep month, then Rains-Rot-the-Ropes month. (The ropes are the ones used to fetter the calves.) Lakota Indians traditionally measured the year in moons: "the Moon when the Grass is Green," "the Moon when the Chokecherries are ripe" (mid-August), "the Moon of the Falling Leaves" (early October), or "the Moon of the Snowblind" in late January. The variations of landscape or animal appearances characterize both time and locale, in vivid contrast to the West's dominant time-measurement, its invariance characterizing neither time nor place.

Other peoples characterize time by starscapes. The Kiwi people of New Guinea date their months according to when various stars slope across the sky and vanish over the horizon. Aboriginal Australians begin the cultivation season when the Pleiades appear. For some Melanesian peoples, the moon's waxing and waning prompts the human actions of farming, hunting or fishing. For the Tukano people of Colombia, the Milky Way is a calendar for most activities throughout the year. (It is the rainbow which represents the Master of Animals' ejaculation.) According to anthropologist Gerardo Reichel-Dolmatoff, the Tukano say that the snakes rise out of the waters to watch the stars in the Milky Way, to find out if the time of fish-runs is near. They have more than twenty alternating "summers" and "winters" in a year, including the Summer of the Caterpillar (November).

Sure-yani Eduardo Poroso, a leader of the indigenous Leco-Aguachile people in Bolivia, comments: "The Leco people don't use the Roman calendar but nature's calendar. You can't use the Roman calendar to know when to fish. It's imposed on us for registering births and for baptisms, but it doesn't function for use in the forests. Bats tell us when to fish; when they fly close over the water." (His own name is an indigenous time-referent; Sure-yani is the name of a "calendar-tree" which lives for three hundred years and is, he says, "how we measure time.")

The Arakmbut people of Peru still use both the sun and moon as watches, comments Matteo Jicca, an Arakmbut leader. "In the past, we never counted beyond twenty. We never counted years, we would not calculate or count. For people, we wouldn't count their ages but their stages of life—childhood or puberty, youth, maturity and age." The changeover happened as a result of white invasions "when we learned to write." He also contrasts the urban clock-time of the colonizers with his peoples' time in nature. "Your people are all planification and punctuality. In the cities, everything has to be at the hour, *punto*, precise. By contrast here in the mountains we give things time, *sin limitar*, without limits."

In the forests of the Andaman Islands in the Indian Ocean, people have a scent-calendar, using the smells of flowers and trees to describe the time of year. The Arakmbut people also use what they call the "calendar of flowers or of fruit." Matteo Jicca refers to the "time of the pijuayo" when the pijuayo fruit is ripe, roughly the second two weeks of March. Another measurement is "when the pineapple is ripe" in October and November. These dates, he points out, vary and are flexible, in contrast to invariant, inflexible, mechanical countings. The Huaorani of Ecuador have a similar calendar, February, for example, being the "season of the palm fruit." Navajo mythology describes how the First Man designed the calendar. First, he drew a diagram in the sand to show the seasons; these were divided into summer and winter and the months were marked and named with their characteristic happening. November, he called the month of Slender Wind; January the month of Snow Crust; February was his month of the Eaglets; April that of the Delicate Leaves and September that of Maturing Vegetation. To each month was then ascribed a "heart" to characterize it further and a "soft feather" attribution which indicated its good fortune. The month of Snow Crust had ice for a heart and the morning star for its soft feather. Another, the month of Large Leaves, had wind for a heart and its soft feather was rain. The dominant Western clock (11:48) or calendar (06/30/05) preoccupied with endless numbers, numbs any feeling for time itself. No heart there, no soft feather.

Everywhere across the world there are different ways of choosing time-periods—and different reasons. Most societies seem to have some form of week, though its number of days could be anything from three to sixteen. The Babylonians designed a seven-day week, which was adopted by Judaeo-Christianity. The Inca used eight-day weeks (and at the end of the week, the king swapped wives). The Maya, for whom time was a deity, considered a 260-day cycle the most important time unit in the calendar, thought to be because it represented the nine-moons-time between conception and childbirth.

Unlike the modern Western model of time, so disembedded from nature that its time is more precise than the earth, to many cultures time can only be embodied in the natural world and in nature's processes. In North American Indian languages, the term "world" also means "year," the turning of one is the turning of the other. Dakota Indians say "the year is a circle around the world" or around the sacred lodge, for the lodge is itself an image of the world. The Yokuts say "the world has passed" which translates as "a year has gone by." To ask "How old are you?" the Guarani-Kaiowa Indians, from Mato Grosso do Sul in Brazil, would ask "How many times has the Guavira flowered in your lifetime?" In Spain, you can ask "How many Aprils do you have?" to ask someone's age—April cherished for the sheer exuberance of its flowering vitality. (The traditional way to ask a person's age in Korea is to ask his or her zodiacal animal.) A surviving Mixtec year glyph describes a "Year Three Flint." For the Aztecs, time unfolded in five successive suns, or aeons, a Sun Four Tiger, a Sun Four Wind, a Sun Four Rain and the Sun of the present called Four Movement. The traditional Chinese calendar is Dragon-yeared,

Horse-yeared and Dog-yeared. The San Bushmen of the Kalahari refer to the wind and insects as the timekeepers, because they are accurate signals as to the age of an animal track. (The more affected the track is by wind and insects, the older it is.)

There is in the Micmac language—and in most of the Algonquin languages—no word for "time" in an absolute sense. There are words for day and night, sunset and sunrise, the cycle of the year and of the moon, but no "time" as numbered measurement. There are no words for hour, minute or second though there is "now"—*neegeh*. Rastafarians make a distinction between time and clocks, so, asked "What's the time?" may well reply "Clocktime is half past seven." A San Bushman of the Kalahari, asked how old he was, replied: "I am forty or forty-five, maybe. I have no need for years."

In Burundi, time is described—characterized rather than counted—the time of a pitch-black night is called a who-are-you-night, because it is too dark to see a face and you have to ask the identity of people you meet. The Kaluli people of Papua New Guinea have a clock of birds; women say the early morning calls of the brown oriole and New Guinea friarbird and the hooded butcher-bird tell the children to wake up, and later in the day, the birds' afternoon calls tell the children to return to their families. For the Nuer of the Sudan, there was the cattle clock; counted by the tasks of milking, pasturing or moving the cows from byre to kraal. In Rajasthan, in India, the moment when the herds return at evening is called "cattle-dust time." Cow-time is local, social and embedded in nature's processes. Whereas clock-time is global, applicable anywhere, cow-time is local to the very udder-hour.

For North American Indians, there is a reciprocity between humans and time; F. David Peat, a Western physicist who has studied indigenous science, says that for them the concept of time is dynamic and "animate" and there is a sense of deep exchange between humans and time. "Time is not independent of us nor of the rest of nature. Time is addressed in ceremonies and a people's relationship with its movement must be renewed." Time is alive.

Bees can tell the time and be punctual to a daily eating rhythm because they have an internal clock. They can stay punctual to that clock even when moved across continents; if that happens, they, in effect, get jet lag. Co-parenting Emperor penguins have worked out timing to a tee. The mother lays an egg, then leaves to hunt food. The father sits on the egg for two months without eating until the egg hatches, then he feeds the chick from "milk sacks" which run out after ten days, but at that exact time there's a dot on the horizon; the mother returning. In *A Sand County Almanac*, Aldo Leopold writes of the importance of timing for geese returning in March: "A cardinal, whistling spring to a thaw but later finding himself mistaken, can retrieve his error by resuming his winter silence. A chipmunk, emerging for a sunbath but finding a blizzard, has only to go back to bed. But a migrating goose, staking two hundred miles of black night on the chance of finding a hole in the lake, has no easy chance for retreat. His arrival carries the conviction of a prophet who has burned his bridges."

Nature's multifarious clocks are evident everywhere; in daily, or circadian (approximately one day) or in monthly or yearly cycles. Some much-more-than-circadian cicadas live out a longer rhythm; as nymphs they remain underground for seventeen years. Then they all come out at once. In Ohio, thousands are recorded as coming out of their pupal skins, chirping, crawling, climbing, mating and dying. Then the seventeen-year cycle starts again.

But perhaps nowhere is there a more fascinating complex of time than in the seas. Cicero recorded that the flesh of oysters at sea waxed and waned with the full moon (and so, women know, does the human "oyster"). Seahorses mate at full moon. (And the male ones, rather delightfully, get pregnant.) Mussels keep in synchrony with the sea tides of the place where they were born and will adhere to this even when moved to still water. The palolo-worm in the Pacific and Atlantic oceans reproduces only during the neap tides of the moon's last quarter in October and November.

Humanity has long known—and used—the sea's times. In some places, such as Puget Sound near Seattle, the sea's clock is one of almost unfathomable complexity, complicated to the point of eccentricity. But Native Americans, having learned to tell its time, could exploit its ferocious and quick tides, using it for com-munal activities and for collecting food—the gathering of clans and clams alike. South Pacific islanders can predict to the day and hour the beginning of the annual spawning runs of many fish. Meanwhile on the Pacific coast of Canada, a Haida woman, Gwaganad, in speaking of the Haida lifestyle threatened by log-ging in Gwaii Haanas (South Moresby), tells of part of her "job, her purpose" being to gather herring roe. "In the spring the her-ring come and they spawn on kelp . . . It's a spiritual thing that happens . . . I want to share what goes on in my spiritual self, in my body, come February. And I feel it's an important point . . . My body feels that it's time to spawn. It gets ready in February. I get a longing to be on the sea. I constantly watch the ocean sur-rounding the island where the herring spawn. My body is kind of on edge in anticipation. Finally the day comes when it spawns. The water gets all milky around it . . . this wonderful feeling on the day that it happens, the excitement, the relief that the herring did indeed come this year. And you don't quite feel complete un-til you are right out on the ocean with your hands in the water harvesting the kelp, the roe on kelp, and then your body feels right. That cycle is complete."

The sea, clock of ages, is full of time. In the tide's ebb and flow the sense of the moment is critical, but it is the coasts which are affected by tides, not the ocean depths, so while the sea, at its shoreline, represents the *now* of events, yet the paradox of the ocean is that in its depths it is a symbol of *eternity*. (Byron called the sea "the image of eternity.") The everlasting consolation of the sea is not *all will be well,* but *all will endure.* To Western scientists, the sea is the source of life. In Taoist thought, similarly, the ocean is equated with the Tao, the primordial and inexhaustible source, "informing at creation without being exhausted." Jainist thought of the sixth century B.C.E. describes an "ocean of years" being one hundred million times one hundred million palyas. Each palya is a period of countless years. Otis Redding picked the right place, "sittin' at the dock of the bay, wastin' ti-ai-ai-ime," for the sea is creator of endless hours of time. And this is one reason why polluted dead seas are so shocking, for it suggests the poisoning of both the actual sea and the conceptual source of time itself.

Today, the ocean's clock is threatened; in the flood tides, freak storm surges and predicted rising sea levels which are the result of climate change. With our increasingly profligate use of central heating and air-conditioning, nature's time and climates are controlled, but then these very systems provoke climate change—in a self-fulfilling profligacy. This control is "imposing our wound-up present on the slow time of nature," according to Jean Chesneaux, Sorbonne professor and author of *Brave Modern World.*

Affinities between the idea of the sea and of time run deep and wide. Time and tide which wait for no man have drawn the human mind from ages past; in English the word *tide* can be used for both the sea's tides and certain specific times; noontide, eventide, Whitsuntide or Eastertide. The word *tide* is etymologically related to the word *time* in Old English. "Current" refers to both time and tide.

In a metaphoric sense, probably the single most common image for time is a river running to the sea. In Micmac society, with no word for time, the river is an image for the flow of happenings. The Pirá-paraná Indians of northwest Amazonia show the cycles of life and death as a huge circular river. The ancient

Greeks identified time with Oceanos the divine river, which circles the world. Time is "like an ever flowing stream," according to a Christian hymn. Matthew Arnold referred to time as a river, "the natural and necessary stream leading to the future." "Time glides by with constant movement, not unlike a stream, for neither can a stream stay its course, nor can the fleeting hour," wrote Ovid in *Metamorphoses* XV. The thirteenth-century Zen master Dogen said: "The time-being has the quality of flowing."

Ancient Indian thought describes time as a flowing of water and compares it to an ocean. The Hindi word *ab* means both "water" and "now." The Sanskrit text, the *Atharva Veda*, meanwhile, speaks of a full or overflowing vessel placed above time. The vessel is itself timeless but time is forever replenished from that source. It is a rich image of both time's flowing liquid nature and of its characteristic fullness, full as an ocean.

In a bodily sense, the human source of (life)time is the oystery vulva, which can taste so remarkably like the sea; while the womb is tidal in its ebb and flow, and as responsive to the moon as the sea's tide itself. A woman with child has her "waters," and pregnancy can make women feel, they say, oceanic.

In a historical sense, time and the sea are also linked: the great breakthrough of timekeeping—making clockwork of sufficient accuracy to discover longitude—was done in order to gain control of the seas, to colonize, if you like, the wildness of ocean-time, a feat which led to Britons ruling the waves and, through this, ruling both empires of land and empires of time, for it was due to Britain's maritime supremacy that Greenwich was accepted worldwide as the zero meridian. Ruling the seas meant ruling the standard of world-time—*the* time which is so highly political and so seldom recognized as such.

So, after all these different times, see how false is the ideology of Western imperialist time, declaring itself *the* time, when it is just one of thousands upon thousands, see how Greenwich Mean Time is mean indeed, as indifferent as it is banal and as cruel as either.

The whole way time is described is ideological, if invisibly so (and, as a later chapter shows, time has always been tied to power). Today, the West's numerical time, its obsessive dividing, atomizing, counting and accounting of time is ideologically fitting to the industrial and post-industrial age, which began emptying time of any distinction or variety with Newton's "absolute, true and mathematical time," and which learned to say time-is-money with the miserable Franklin, so today the fullness of time is ever emptied—of its grace and generosity—by today's mean and graceless accountants-of-time. In femtoseconds and cesium atoms, modernity's time is divided but not distinguished. The digital clock fragments time, splits the present off from the past or future, an instrument of fission not fusion, for this age of fission. The digital stutter repippipppippetitively reproduces itself. In contrast to the Native American belief that time is alive, the digital watch doesn't animate the present but kills it with chronic momentism, emptying time of character as it atomizes it.

The atom was once considered unsplittable (derived from the Greek for indivisible). It is so appropriate that as this was the age which split the atom so this is the age which uses the atomic clock to split the split second. There are social rhymes with this; the fractured, split communities of today and the little social division that happens every time you glance at your watch. Watch how.

Take one common example; someone is chatting to you and you surreptitiously look at your watch. Why the stealth? Because the glance is a faintly unkind interruption, a small social disjunction. Take another example. In New York, Bilbao, Seoul, London, Munich, or Paris, if you want to catch a bus, you check your watch and a printed timetable, rather than ask a person. The worn watch isolates its wearer in social fission. In India, say, if you want to catch a bus and you start scrutinizing watches against timetables on pillars, people will laugh. (I have. They did.) You have to ask a person in a humble piece of social fusion. No Indian bus driver (and purely for the fun of digression, do you know the Hindi for bus driver is *bas Draaiivar*?) would be seen dead being ordered to leave by scrappy bits of paper glued up seven years ago. Buses go and people know when, for here time still resides in the human, not in the abstract, paper world. The outrage felt by those used to living in human-time forced to live by an abstract (and foreign) clock was graphically expressed in a letter from an Indian man to the white superintendent of the Indian railways in 1909: "Dear Sir: I am arrive by passenger train at Ahmedpur station and my belly is too much swelling with jackfruit. I am therefore went to privy. Just as I doing the nuisance that guard making whistle blow for train to go off and I am running with lotah [waterpot] in one hand and dhoti in the next when I am fall over and expose all my shocking to man, female, women on platform. I am get leaved at Ahmedpur station. This too much bad if passenger go to make dung that dam guard no wait train five minutes for him. I am therefore pray your honour to make big fine on that guard for public sake."

In a more conceptual sense, modernity's obsessive measurements of time alienate you from time itself; the divisions of clock-time divide you from the actual experience of time as measured, say, by your hunger, or by the buzz at a café. Listening to the pips on the radio, you can't listen to time *en ciel*, to a skylark's jazz riff. Measuring the moment prevents you from feeling its heart or being tickled by its soft feather. Tracking time stops you in your tracks.

Humanity's relationship to time is poignantly reflexive—

being too "exact" about time is exacting, stressful, for people. Sociologist Barbara Adam in *Timewatch* notes how clock-based urgency results in high blood pressure, a battered immune system and ulcers. Too-frequent deadline times, studies (such as those of Dr. Meyer Friedman) show, can distort one's perception of time and can disturb the pace of the human heart, which can in turn cause heart disease and death. *Dead*lines, indeed. Dr. Friedman's "Type A" personalities epitomize the reflexivity between humans and time. Type A traits include impatience, aggression and a harried sense of urgency and Type A personalities appear to be in a chronic battle with time and life itself: victims of their own "hurry sickness," they are more likely to suffer heart disease. Ruled by measuring momentism, humanity lives at the mercy of its own invention, machine become monster. The faster the watched clock moves, the slower the watcher can. A watched clock never strikes—but the pause is really in the watcher, motionless. The stopwatch *stops* the *watcher*, halted in hypnotized inaction. Are they telling the time or is time telling them, ticking them off with a tedious, relentless, "now now"?

For time does not mean watches, clocks or the oscillations of cesium atoms, time is not found in digital pips or paper calendars, time is not in pendulums or in chronometers; the clock is not a synonym for time but the opposite of time. The West's obsessive time-measurement has gone hypertelic, beyond the point of usefulness, and the clock of the present is not the realization of time, but its betrayal. Society begins to think in the forms it has structured for itself, linear, artificial, over-fragmented, modeling itself in the image of its machinery. Today's timekeeping pretends it is describing time ever more accurately; what it really describes is modernity, in a telling self-portrait. Modernity ascribes to time a driving, rigidly linear, impersonal, coercive and dominating character, overcrowded and overwound, which harries its victims, people. You hear the screech, time is running out—as if that were time's fault—but it is modern society itself which is overscheduled and crushingly domineering in its timings.

Clocks, incidentally, need not be invariant or mechanical; in 1751 Carl Linnaeus, the Swedish naturalist, designed a floral clock of flower-hours, so the time of day was shown by the staggered blooming of various flowers, the spotted cat's ear at six in the morning, the passion flower at noon and the evening primrose—naturally—in the evening. Essayist William Hazlitt had a beautiful summer-generous inscription carved on his sundial; *Horas Non Numero Nisi Serenas*—I count no hours but those which are serene. Before GMT, nine of the clock wasn't nine of all clocks, but nine of an agreed clock, a localized clock, possibly even a named clock. (It is hard to imagine a digital clock being named and even their faces are *feature*less. In them is neither day nor night, summer nor winter.)

In the West, since the early years of this century, time has been increasingly homogenized. The mono-time of Greenwich *mean* time offers no sense of time's generous variety, nor any intrinsic character or color to time, it is an artificial construction and, more than anything, modernity's time—the global present—is increasingly standardized, increasingly the *same*.

It is mid-morning, mid-May. It is always mid-morning, mid-May, in the shops and offices of the present; the light is the same and the temperature is the same. It is Same o'clock in the month of Same; modernity's time has nothing to do with "real" time—nature's time. It is instead a manufactured time: synthetic time. Jack Frost is dreadfully unpunctual when it comes to showing up for Christmas; fake frost is far more punctually dependable for the window display of modernity. Time is divorced from nature as plastic plants pretend to a perennial summer they never knew. Year-round Astroturf tramples on real spring's chilly-kneed grass seedlings, and neon knifes night.

Food products were once obstinate markers of time: the Nut month and the Watermelon month of the Natchez tribes, for instance, or vegetables in season. Now there need be no such thing as seasonal produce. It's always nut month. Nuts in June and watermelons in November. The new potato has ceased to exist since imports fill the shelves all year round and what is always new is never new. A recipe for *summer* pudding begins "seasonality not being what it used to be, there is no reason why you shouldn't make this pudding . . . all year round." The rose growing season is a wistful anachronism since roses can be bought at any time. The Japanese symbol of beauty, the cherry blossom, is seasonality at its most sublime, representing the aesthetic of the evanescent, but when cherries are forced to blossom all the time, the fragile beauty of fleeting time is void of meaning. To live in a synthetic ever-present present is to live not in the fullness but in the *emptiness* of time.

Western modernity's attitude to time mirrors its attitude to place and as a key phenomenon of place today is homogenization, the erosion of place-distinctiveness to a global suburbia, so likewise time is increasingly made an indifferent, indistinct suburban same-time. Appropriately, the greater the citification, the more this sameness of time is apparent; urban street lamps turn night to day and city shops stay open later into the night than rural shops. The very seasons impinge less on people in cities: the difference between a new moon and full, powerful in a landscape, is all but immaterial in a streetscape. While nature's time can be specific to a cow's udder-hour, or to a certain frozen creek at new moon: cities, ever more independent of nature, use the unspecific global rhythm of the artificial clock. Jean Chesneaux picks out cities as "modernity's special places," the modernity which "ties itself to the perpetuation of its own present." Same time, endlessly repeated. Same o'clock.

The forest is the symbolic opposite of the city: according to

Shakespeare, "there's no clock in the forest." But the forest is one great big gong of time, as the Karen knew. Across the world, all nature is in time and time is in it. Every apple tree hears the pips, every pip in every pippin picking the moment to split. Birds judge the calendar for their migrations with annually remarkable accuracy and in flight their timing and synchronization is a miracle made ordinary. Time is everywhere in nature. In urbanized life, clocks are needed precisely because there is no other way of telling the time. But while nature knows a million varieties of time, the clock of modernity knows only one. The same one. Everywhere.

While once, you could say that time was so local that for every *genius loci,* a spirit of specific place, there was a *genius temporis,* a spirit of specific time, the history of Western timekeeping has been one of standardization and of globalization. The march toward mono-time began slowly, a tread here—from 1840, railways in Britain required a standardization of time—a tread there—in 1880 London Time was decreed by law to be *the* time for the whole country. A step here—in 1883, the U.S. standardized timekeeping for their railways—and a step there—in 1884 Greenwich was made the zero meridian and the global day of twenty-four hours began, a day made of the same hours, irrespective of local cultures' clocks. The French opposed it until, at the 1912 International Conference on Time in Paris, they accepted that though the meridian would be English, world-time would first pip in France. 1912 was a fateful year for time.

For when historians discuss the moment when the gradual spread of one world-time-reckoning became an actual *experience* of instantaneous time, the moment is midnight on April 14, 1912. The distress signal sent from the sinking *Titanic.* The news flashed around the world via the telegraph. People referred to "a new sense of world unity" and what is now called the creation of the "global present." Its staging was so appropriate; taking place not on land but at sea; the sea, source of time and of such intricate varieties of time, was to be the setting for the first drama in global time. And when the actual *Titanic*—the ship—sank, it launched a metaphoric *Titanic*; huge, monolithic, global time.

The media (newspapers, telephones, radio as well as the telegraph) buzzed with the news, blaming the tragedy, incidentally, on an obsession with timekeeping over safety. Why did the accident happen? asked the *Daily Herald*. Because "liners must not get behind their scheduled time. If humanly possible, they must get in front of it."

But the media itself was implicated not only in the creation of this global present but also in the portrayal of time. (A 1914 editorial in *Paris-Midi* described a daily paper's headlines as "simultaneous poetry.") The association between the media and present-time is so deep that almost every part of the media is named with a time-referent; *Paris-Midi* itself, *Time* magazine, the *New York Times*, *Tiempo* and *Epoca* in Spain, *Die Zeit* (the Time), *Asahi Shimbun* (the Morning Sun newspaper in Japan), *Time Out*, the *Daily Telegraph*, the *Times of India*, *Today*, the *Today Programme*, *The Atlantic Monthly*; Barcelona's newspaper suggests it is "ahead" of time, titling itself *La Vanguardia*. The earliest portrayal of time by the global media was novel: the use of the telegraph, for instance, changed newspapers; once places of cultural *memory* (actually called news *history* in the nineteenth century), they became, in the early twentieth century, more focused on *change*. But more than novel the portrayal of time was, albeit invisibly, very political. The new global media technology, because of its mechanics, favored the short over the long. Information became broken and fragmented rather than continuous. Short facts. Bits. And the things which were above all most easily transmitted were the prices, market news and all the information which commerce and industry wanted—and which therefore assumed a further prominence. All this helped to paint time itself anew, in a ruptured-repetition, quick, short and standardized. And ever more intimately equated with money.

Today, the media defines, to a great extent, what "the present" is, and characterizes the "now." The BBC's rolling news service is advertised by a trailer: "This is the Now O'Clock News from the BBC," instead of the familiar "Nine O'Clock News." Now is always news. (Contrast this with a number of years, such as 667 and 668, left entirely blank in the *Annales Cumbriae* as if to say "nothing happened this year," a phrase allegedly written one year by a decidedly underwhelmed Welsh chronicler.) With almost constant time signals on CNN the present is announced almost *as* a news item, this ever-present present, as the TV *presenter* constantly reiterates. The effect can become hypnotic. Modernity surrounds itself with scrolling TV, now, *now*, imploding headlines, today, *today*, pages of endless newsprint rolling sideways into each other's gutters, the text's meaning escaping into a blank postmodern margin, guttering the spectator's attention, too, in its fold. Media-time demands your attention to the extent that real-time in your material world can seem immaterial. TV-time in the Gulf War seemed more real than real-time.

8:45 A.M., September 11, 2001. Another date in world-time, the first hijacked jet hits the north tower of the World Trade Center. Television cameras are immediately trained on the Twin Towers. It was another experience of the global present—like the *Titanic*. Like the *Titanic*, the World Trade Center was considered invulnerable, like the *Titanic* its construction was an expression of national superbia, of superiority, the towering colossus. But this global present, unlike the sinking of the *Titanic*, was watched as it happened. People all over the world phoned each other, told each other to switch on the TV only to see the second jet hit the south tower, at 9:03, the moment when what appeared accident was confirmed deliberate. This most momentous televisual event was broadcast globally: repeated, over and over, fragmented and repeated, fragmented and re-repeated. The hub of world trade and finance which depended on the instantaneous *now* of market news and prices, was itself the subject of the instantaneous media *now.* Time was used to magnify the tragedy: The Day That Changed The World Forever, said one newspaper. The Day The Earth Stood Still, said another. So much an example of the global present, the atrocity was named by its date: September 11. The *now* of global finance and media was rendered in almost sinister understatement as, shortly after the towers fell, on the website of the World Trade Center, its firm listing was, it said, "outdated." Times Square was where many New Yorkers gathered to hear the news, echoing history; Times Square not only being the home of the *New York Times,* but also the site of the first on-the-spot wireless dispatch, an account of fighting off Port Arthur in China.

CNN ran a constant headline three days later: "America's New War"—preferring news to history, effects to causes. The New War Now-News went global as the ninety minutes of "the most dramatic TV pictures ever" played and re-played in re-creation of the global moment. The media's soundbite culture was manifest. One woman was being filmed as she watched the second tower fall: "Oh my God, my brother-in-law's in that building." "That's a soundbite," said the Channel 11 reporter to her producer. "Hang on to that soundbite." True to the pattern of the media reporting only short facts and bits, one-off moments, the Manhattan attack was fragmented from its causes. Treated as a unique atrocity. Isolated. Without cause. An atrocity it was. Unique, isolated and causeless it wasn't.

But back to those early years of the twentieth century, when something so exceptional was happening to time. Keep watching Paris. 1913. Just a year after the *Titanic* launched the global present, at ten in the morning on the first of July, at the Eiffel Tower, the first worldwide time signal went pip. (Much to the chagrin, one presumes, of the British, who had led the race in time-measurement for so long and who were now literally "pipped at the post.") And at that moment with a wireless signal traveling at the speed of light, one time was pronounced for the whole world. Essential for global business and travel, this meant Western (Christianized) time was made a world standard.

Any exaggerated tendency in society will (as ancient Chinese philosophy knew so well) throw up its complementary opposite and at that point in history, two remarkable aspects of time came into being. Keep your eye on 1913. For that was also the year Henry Ford's moving assembly line first swept into action, cutting the time taken to make a car from fourteen hours to just two. This production-line-time would be used later as a model by ideologues of both the extreme right and left. And just at that point in Paris, when the global present thinned time to a radio beeeep, just when time had never been so public nor spread so thinly, simultaneously composers and writers gave time a fullness, a thickness and texture it had never had before. Cinema, too, used slow-motion to vary time and "stills" to privilege one moment forever, as in the 1913 film *Atlantis,* inspired by the sinking of the *Titanic,* its haunting stills freezing a single second. The Cubism of Picasso, meanwhile, in this period, expanded the dimension of time. As it shattered the idea of linear space in traditional perspective, so it broke with the traditional sense of time in that an object was seen not as a camera might see it but as the mind knows it; different aspects of one object revealed simultaneously, the present broadened, thickened and filled out by the subjective mind.

In Paris, in May 1913, Stravinsky's *Rite of Spring* was first performed—an example of simultaneity in music due to its tri-tonal harmonies. At this moment in history, something strange happened to simultaneity. It both appeared to happen and was denied. Simultaneously. On the one hand, there was Stravinsky and cinematography, plus the increasing number of people experiencing simultaneity through, say, the wireless. On the other hand, Einstein. In his 1905 and 1916 theories of relativity, he argued there was no such thing as one simultaneous time. By contrast, "every frame of reference, or moving body, has its own time."

Stravinsky's *Rite of Spring* spoke profoundly of *nature's* time, of pagan time, in the very year, 1913, which saw Paris's *urban* time made global. It describes a pagan ritual dance to induce spring to come and time to move on. (The implacably Christian Western calendar had for centuries been usurping the time of nature, of pagan time.) The second act is "The Great Sacrifice," a dance to death; a human sacrifice is required for nature's time.

The relation between time and sacrifice is perhaps a little-acknowledged human universal. According to ancient Indian tradition, if the priest did not offer up the sacrifice of fire every morning, the sun would not rise. Almost all midwinter festivals past and present involve some kind of sacrifice: as J.G. Frazer argued in *The Golden Bough*, the young god's sacrifice pushed the year onwards and in the Saturnalia, the figure of Saturn was killed in sacrifice. (Saturn was associated with Chronos—god of time.) Pagan tradition speaks of a sacrificial battle at midwinter between the oak king and the holly king to turn the year forward. The human sacrifices of the Aztecs were performed to strengthen the sun in its journey through the year. The Mexican and Mayan sun god needed human hearts in order for time to move on. In India, the goddess Kali is associated with both time and sacrifice. The connection between time and sacrifice, in the West, was never so horribly thickened as in the First World War when some historians, such as Stephen Kern, argue that seven million people were sacrificed *because of* the manufacture of global time, for the war began as a direct result of the new global present. The instantaneity of the telephone and telegraph left no diplomatic pause for thought, no time for the discreet, virtually Viennese waltz of *temps diplomatique*; time, compressed too much, exploded. After the assassination of Archduke Ferdinand in Sarajevo, Serbian diplomats were given a forty-eight-hour ultimatum to explain the shooting. The Serbs missed the deadline and the killings began. Spengler was to note in 1918 that

> in the classical world, years played no role, in the Indian world, decades scarcely mattered, but here the hour, the minute, even the second is of importance. Neither the Greek nor the Indian could have had any idea of the tragic tension of a historic crisis like that of August 1914 when even moments seemed of overwhelming significance.

Global time began with the tragedy of the *Titanic*. And the first thing it caused was a tragedy. After the attack on the World Trade Center various publications were baying for instant reprisals: the global present demanded immediacy and justice could go "to hell" as the *New York Post* put it.

Stay with Paris, 1913. Just as Stravinsky's *Rite of Spring* was first being performed and just as the Eiffel Tower was sending its first global time signals in July 1913, making time standardized and universalized, public and objective and explicit, so with exquisite timing the first part of Marcel Proust's *À la recherche du temps perdu* was being published from 1913 onwards. While time represented by the Tower was a flip, regular blip, every day with exactly the same duration: time represented by the novel was so various that a day could last 287 pages while some years could slip by without a whisper of record. And time, which in one sense had never been so public, so monolithic, had in another sense, with Proust, never been so private, so unique, so local to the psyche's own hour—all madeleines and murmuring memories of the time of the mind. In the title is time; in the first sentence, time; in the last word, time; through all the volumes, time is the subject;

A SIDEWAYS LOOK AT TIME

26

the longest longing reverie on time ever written, its ending describing humanity

> occupying a place, a very considerable place compared with the restricted one which is allotted to them in space, a place on the contrary prolonged past measure—for simultaneously, like giants plunged into the years, they touch epochs that are immensely far apart, separated by the slow accretion of many, many days—in the dimension of Time.

And so to now. With its dominant ideology, the West declares its time is *the* time. Not so fast. Its dominance is actually far from complete. Its challengers are everywhere. The human mind will always be unclockable, for in one moment it can flit between a pumpkin and a paper mite, can think of chaos theory, breakfast and the two-toed sloth, can shudder at a bad memory and consider the billion-year nature of star-time. All in a second, so ungovernably, so splendidly myriad-momented is the mind, as Proust knew so well. Time, as a later chapter will show, is subtly genderized and while the dominant, linear calendar is masculine, every bleeding woman knows a different female, cyclic time.

Though *the* time has an even, perfectly regular beat, through the length of the human life, time does not move with any such mechanical regularity. While old people sigh over how fast time goes, children are incapable of patience. How long is an hour to a child? Far, far longer than to an adult; asking a small child to wait a few hours for ice cream is like asking yourself to wait a week for a whisky.

Adults, generally, have learned clock-time. Children live in the heart of the ocean of time itself, in an everlasting now. The writer R.K. Narayan describes childhood as "letting the day pass without counting the hours. One existed in eternity." Any pipsqueak of a child can do it. A child's eternal present is present-absorbed, present-spontaneous, present-elastic. Children have a dogged, delicious disrespect for punctuality and an innate dishonoring of the dominant clock of modernity.

Time, it is said, is the most widely used noun in English. Time can mean timing, or ending: there is circadian time, bee-time, cow-time and coconut-time. It can be wasted with Otis and sold with Franklin, it can be wrong and be right, be past or future or present. "What," asked English physicist Michael Faraday, in a letter to Benjamin Abbott in 1812, "is the longest, and the shortest thing in the world: the swiftest, and the most slow: the most divisible and the most extended: the least valued and the most regretted: without which nothing can be done: which devours all that is small: and gives life and spirits to every thing that is great? . . . It is Time." In Stephen Hawking's words, "there is no unique, absolute time."

While the dominant calendar insists on its hegemony, its opposition is inside everyone—you can't beat spring for a sap-riser, in the trunk of the tree and in the trunk of the human body, that time of year, that vivid, verdant moment, the ache of the first smell of the year's first grass, so green it's practically acoustic. What Dylan Thomas knew, "The force that through the green fuse drives the flower / Drives my green age;" is what Henry Thoreau, a man keenly tuned to nature's time, described, positively dizzy with sap rising: "Let us preserve religiously, secure, protect the coincidence of our life with the life of nature . . . My life as essentially belongs to the present as that of a willow tree in the spring. Now, now its catkins expand, its yellow bark shines, its sap flows, now or never you must make whistles of it." (*Journal*, January 26, 1852.) For the Penan people of Borneo, comments Wade Davis, time is not measured in hours, but experience. If a hunting trip is successful, it is considered to have been short, though it might have taken weeks by a Western measurement.

Nature was once the biggest public clock, its rhythms establishing "time communities" of joint activities, a shared observation of a natural event or season, a shared sowing, a shared harvest. Modernity, while losing this marker of common time, has nonetheless created others in its place. Sports events: the Ryder Cup; the Superbowl; the U.S. Open; La Vuelta Ciclista España can be hugely popular, in part because they are shared experiences in a common time. Television, too, characterizing time and season, can replace nature as public clock and calendar. It can honor seasonal difference with its sultry summers of French films and traditional Christmas fare. It can distinguish night and day, with its peak times and its watershed times and with the peppery best of its evening viewing compared to the invalid jelly of daytime telly. Television can offer a sabbatical pulse even in its advertisements; alcohol and pizzas on Saturday evenings and hangover cures on Sundays.

The ancient Greeks had different gods for time's different aspects (including the god of the moment for weeding, the god of the moment of horses panicking, the god of the moment when a party suddenly falls silent). One of the most important was Chronos who gives his name to absolute time, linear, chronological and quantifiable. But the Greeks had another, far more slippery and colorful, god of time, Kairos. Kairos was the god of tim*ing*, of opportunity, of chance and mischance, of different aspects of time, the auspicious and the not-so-auspicious. Time qualitative. If you sleep because the clock tells you it's way past your bedtime, that is chronological time: whereas if you sleep because you're tired, that is kairological time. If you eat biscuits when you're hungry, that is kairological: whereas if you eat by the clock, that is chronological time. (In English there is that quaint-sounding mid-morning meal literally named for the clock: elevenses.) Children, needless to say, live kairologically until winkled out of it. Chronos was considered by the Ancient Greeks, and the modern West, as superior to Kairos. Astrology is time considered kairologically: in Hindu life, for instance, the time of the individual and that of the cosmos are considered inseparable. The traditional zodiacal animals of Korea are also used to name the twelve-hour periods into which the day is divided and impart their characteristics to each period and to those born in that period. While astrologers (and, I'm tempted to say, "Kairopractors" but I won't) see a rainbow of colors in time, the dominant calendar of the West is *strictly magnolia*.

Kairological time has a different sense of movement compared to chronological time. For a rough comparison, contrast an urban with a rural day. In cities, where time is most chronological, you move into the future, facing forwards, your progress through the day is like an arrow while the day of itself "stays still," for time is not given by the day but is man-made, culturally given, and defined by the working-day or rush-hours. In a rural place, days roll over the horizon at you, round and gold as the sun, time moves towards you and is nature-given, defined by sun or stars or rainstorms. In this more kairological time, the future comes towards you (*l'avenir*, in French, expresses that, or "Christmas is coming") and recedes behind you while you may well stay still, standing in the present—the only place which is ever really anyone's to stand in. This experience of time, so unlike the urban, is one reason why the countryside, and access to it, is so vital in overurbanized societies; it offers a kinder time.

Kind but fluky. If chronological time is like worldwide suburbia, kairological time is the *genius loci*, the spirit of that particular moment. Kairological time is far richer—far trickier—a concept; time enlivened and various, time as elastic and fertile as an ovulatory cascade.

The Balinese calendar reflects the distinct and particular character of time: "it doesn't tell what time it is, but rather what kind of time it is," according to Cottle and Klineberg in *The Present of Things Future*. The San Bushmen of the Kalahari do not live by Chronos but by Kairos. They do not schedule when to hunt, but rather "wait for the moment to be lucky," reading and assessing animal patterns, spontaneously and sensitively using the "right" time. Beginning with the question "Why do rituals never begin on time?" Renato Rosaldo writes of the Ilongot people of the Philippines and their sense of timing as variable and indeterminate, optional and unpredictable, and considers it creates a situation where creativity and improvisation, flexibility, fluidity and responsiveness can flourish. "Social unpredictability has its distinctive tempo, and it permits people to develop timing, coordination, and a knack for responding to contingencies. These qualities constitute social grace, which in turn enables an attentive and gifted person to enjoy and be effective in the interpersonal politics of everyday life." Their sense of time emphasizes "the tempo and rhythms that shape the dance of life."

Writers on hunter–gatherer life have evoked a subtle, graceful picture of kairological time. Elman Service comments on the unique, sporadic quality of hunter-gatherer time, rather than the repetitive, numbered clock-time and Hugh Brody, in *Maps & Dreams*, writes of the way hunting decisions are taken among the Athapaskan people of British Columbia; "some of the most important variables are subtle, elusive, and extremely hard or impossible to assess with finality . . . To make a good, wise, sensible hunting choice is to accept the interconnection of all possible factors . . . Planning, as other cultures understand the notion, is at odds with this kind of sensitivity and would confound such flexibility. The hunter, alive to constant movements of nature, spirits, and human moods, maintains a way of doing things that repudiates a firm plan and any precise or specified understanding with others of what he is going to do. His course of action is not, must not be, a matter of predetermination."

In kairological time you have the cursive, oceanic ebb and flow of thought, music, images and the moment of sudden artistic insight. Not surprisingly, artists of all colors have long contested the hegemony of dominant, chronological clock-time. Take Messiaen's 1960 orchestral piece *Chronocromie* (Time Coloring), for instance. Toni Morrison comments, "Time is layered, not linear, because that is the mind's experience." Virginia Woolf, whose characters move in an element of time rather than space, was sensitive to the difference between clock-time and time itself and also aware of the watery nature of time, its ocean and river-symbols. In her appropriately waterful novel *The Waves*, one character comments on the exterior experience of clock-time that

> it is a mistake, this extreme precision, this orderly and military progress; a convenience, a lie. There is always deep below it, even when we arrive punctually at the appointed time with our white waistcoats and polite formalities, a rushing stream of broken dreams, nursery rhymes, street cries, half-finished sentences and sights—elm trees, willow trees, gardeners sweeping, women writing—that rise and sink . . .

James Joyce's *Ulysses* and *Finnegans Wake* are full-fathom fantasies of time, the former expanding a day to a book and the latter mining the moment of language by splitting the very syllable in puns. Since in the human mind time has such affinities with water, it is hardly surprising that work so connected to time should be so liquid, so riverine. *Finnegans Wake* begins and ends with the archetypal river which is time itself running to the seas: the book's last sentence runs back into the first, like the cycle of rain to river to sea. Round, wet time, in opposition to the dominant, linear, dry time of clocks and calendars.

Recently, Dublin authorities—who did not understand their Joyce—sank a digital clock, which was supposed to count down to the millennium, underwater in the river Liffey. The river, with a far deeper sense of the fullness of time than the authorities, rejected this digital nought, and the clock broke down repeatedly. For if by worldwide imagery time is a river then the clock is merely a mechanical pretender.

But when modern society, in its valuations of time and money, in its media and in its neuroses, in its gridded screen diaries, on the radio, in its financial centers and its mind-set, is so preoccupied with the now: what is it anyway? What is this "present moment" which atomic clocks (don't) define, but which children are born knowing? "What is love?" asked Shakespeare, "'tis not hereafter; *Present* mirth hath *present* laughter." It's an unparalleled expression of the fullness of the present, the now. "Only a man who lives not in time but in the present is happy," commented Wittgenstein. *Vive al dia*, the Spanish say: seize the moment or live up today. Eastern religions speak of using the *now* in the fullest sense, as the gateway to "eternity." The true Sufi is called the "son of the moment." According to Zen master Dogen, "It is believed by most that time passes; in actual fact it stays where it is." Korean-American writer Younghill Kang in *East Goes West*, 1937, writes: "And speaking with an Asian's natural bias, it seems to me it is wrong to say, time passes. Time never passes. We say that it does, as long as we have a clock to calculate it for us. The two hands go, the iron tongue tells hours, we sense the experience of our own duration . . . we are illusioned. It is not time that passes, but ourselves. Time is always there . . . as long as there is life to use it. Only if no life existed, there would be no time. Time was because life was . . ."

"No day and night here, mate. No asleep, no awake. No

dreams, no reality. No past, no future. It's a continuum: Huaorani time. It's all right now," Joe Kane quotes in *Savages*.

Czech poet Miroslav Holub recounts his pleasure on finding that, according to experimental psychology, the present moment lasts three seconds in the human mind. "In this sense, our ego lasts three seconds. Everything else is either hope or an embarrassing incident. Usually both."

The Science section of the *New York Times* recently ran a long article on the idea that time could merely be a "psychological illusion important only to humans, not to physics." And physicist Julian Barbour has suggested time is nothing more than a product of human perception. There is no past and no future. Time and motion are nothing more than illusions. Every moment exists forever: and there are numberless Nows, all of which last forever. "Each instant we live is, in essence, eternal." Interestingly this seems presaged by Dogen when he says: "In this world there are millions of objects and each one is, respectively, the entire world . . . As there is no other time than this, every time-being is the whole of time: one blade of grass, every single object is time. Each point of time includes every being and every world." And he also says: "In essence, all things in the entire world are linked with one another as moments."

Now is just a little, bare, empty word in English, just a scrap between an ancient buried past and a starry future. In French, though, you have *maintenant*—richly expressively, time now, at hand, meaning etymologically "held in the hand." The essayist Montaigne embodies a full-to-the-brim sense of now, handfuls of *maintenant*. He had an almost unrivaled sense of his life leading up to the fullness of the present, with a depth and a rounding out to eternity. To fill out the now, he would "remplir son maintenant."

William Blake's image, "Hold Infinity in the palm of your hand And Eternity in an hour," also illustrates the fullness of now so present to the senses—the moment "at hand." Blake, who understood time's ocean-fullness, made a savage engraving of Newton, responsible for such rigid and false definitions of time. Newton is seated on a rock in the ocean, with a pair of dividers. He sees nothing of the ocean, of time itself, he merely stares in blind fixity, the dry-eyed divider, dividing time, into mere mathematical ratios and at a stroke emptying it of quality and distinctiveness.

Newton was the expert who defined time. Benjamin Franklin's definition that time is money, has been bought—on tick-tock—ever since. From the first calendar-makers to today's cesium-atomists, we have been persuaded that experts know the time, but the result is that though modernity knows *the* time down to its silliest split second, it does not know time at all, does not know time as if people mattered, or as if time itself did.

The West has been lied to about time for too long: the mean and nasty definitions offered by Franklin or Newton or the National Bureau of Standards, the dry little sums in sand, the caesium accountants' gridded clocks. Dürer's gridded screen but not the nude behind it. These definitions know nothing about time, nothing about the present, the now.

The French philosopher Henri Bergson—who greatly influenced Proust—understood the sublime importance of the present moment ("Time is creation or it is nothing at all") and lived durations are not simply intervals but are the "very stuff of reality." The sense of *durée* (duration) is to him a source of freedom. Protesting against the fragmentation and repressiveness of mechanistic time, he argued for a qualitative sense of time and in watery images, he defines duration as something to be plunged into, and time, he argued, was an indivisible *flow* of experience—not for him the nanosecond. Jean-Jacques Rousseau had a similar experience of the present moment on an island—surrounded by water, by the oceans of time. ("I have always loved the water passionately," he wrote.) He described his waterful experience of time, by a lake or by a brook, in a river-reverie thus: a state "in which the present lasts forever without, however, making its duration noticed and without any trace of time's passage." Rousseau, detester of commanding clocks, lover of freedom, water and of time itself, knowing that the clock is the opposite of time, refused to wear a watch. In "Tiempo sin tiempo," Uruguayan poet Mario Benedetti writes of his yearning for unclocked, unhasty watery time, "tiempo sin recato y sin reloj" (time without restraint and without a watch) "tiempo para chapotear unas horas en la vida" (time to splash about a few hours in my life).

Modernity's definitions of time, time-accountancy, the digital clock with its flip blip for the flitting split second—the same split second—is a thing of no moment. When it comes to defining time, only the oceanic need apply. Time is simply too exquisite to be entrusted to anyone else. Only the oceanic, the Montaignes or Joyces, Shakespeares or Rousseaus, Eastern philosophers or children. They know their *now*, they know the really wild vibe of the present is this; *now* is the only time when the moment can meet the eternal—and they know that moment is momentous. Now is "the instant," etymologically "standing on" the shoreline of time, one toe in the tidal present, one toe in the ocean's deep—and full—forever, the poise of pure potential. Here, with the child on the shore, at the brinkful rim of events—where the Moment is not the opposite of the Eternal but its only possible realization—is where time happens and every watch which thinks it tells the time should be taken off, flung into the sea, into the ocean, into the waters of time itself.

Drown your watch.

2 ◦ F.FWD.

THE TROUSER-ARROW OF SPEED

It is better to have loafed and lost than never to have loafed at all.

—JAMES THURBER

Take the cow. It can be hard to understand the deification of the cow in Indian villages. But pause. In westernized Delhi or Bombay, amid the fizzing pandemonium of the fast lane, watch the awesome cow, in awesome slowness, chew. Then you know.

Speed is something of a holy cow to modern westernized cultures. On the international foreign exchange markets, up to $290 million can be turned over in little more than a minute. News media can communicate events all but instantaneously. Computers can perform 307 gigaflops per second. Speed hustles everything, from microwave ovens, fast food, Polaroid cameras and the quickie divorce to the Western cow itself, bred to accelerate milk yields. In Delaware, speed is used to celebrate Halloween in Punkin Chunkin, a race using superguns to fire pumpkins faster than anyone else. In Florida, Texas and California you will find drive-thru funeral parlors, the corpse behind a glass screen for you to drive past, so you can view the dead faster—a phenomenon which gives a whole new meaning to the phrase "the quick and the dead." (The phrase "drive-thru" is itself a speeded-up term.) In New Jersey you will find a church which offers a fast-track prayer service, with a speeded-up sermon, quick hymn and a rapid prayer, aiming to run at less than twenty-two minutes.

Professional drivers at Britain's Brands Hatch race course articulate one attraction of speed: control. "Being on the ragged edge of control is exciting because you're not far away from the ultimate of not being in control." In racing terms, this is never truer than at the approach to corners, when the G-force of gravity seems to swerve, frighteningly, right at you.

The attraction of speed is only partly the exhilaration of acceleration; it has much to do with competition, with overtaking. One Californian management consultant, an expert on time-based competition, says "Be fast or be last." (You snooze—you lose.) The thrill is not in going fast, but in going faster than the rest. This excitement is not recognized by all societies; to the Kabyle people of Algeria, and many others, speed is considered both indecorous and demonically overcompetitive. (The Kabyle refer to the clock as the "devil's mill.") One anthropologist describes organizing sports for Australian Aboriginal children, but the children so disliked "disgracing" others by outrunning them that the faster runners would deliberately slow down for the others. In Brazil, the Xavante people have a ritual which involves two sets of people carrying two heavy logs, looking, to Western eyes, much like a race: however if one group falls behind, the other will slow down for them. To ask who "won" is a baffling question to the Xavante; for this is not a race but an act of beauty, an aesthetic event with no competitive component. You cannot ask who won a ballet; nor can you here. No overtaking.

But in the West overtaking is a cultural emblem. In global financial terms, the kick is not just for a company to be wealthy, but to be wealthier than its competitors, streamlined, like a car, to overtake. Products, too, are designed to overtake, to supersede previous models, and to do so more and more quickly. In the recording industry, 78s were in pole position for 61 years, LPs for 26 years, cassettes for seven, and CDs, so far, for eight. But the overtaker, the new mini-disc, is already revving up to pass them (marketers hope) fast. Language, too, is driven faster and faster, markets become supermarkets become hypermarkets. From text to hypertext, words are pressed not to supersede but to hypersede themselves.

According to a Harvard researcher, the average block of un-interrupted speech on television by presidential candidates in 1968 was 42.3 seconds. By 1988 it was 9.8 seconds. Robinson's *Americans' Use of Time Project* showed that in 1965, twenty-five percent said they felt hurried. By 1975, twenty-eight percent said so. By 1985, thirty-two percent felt rushed all the time and in 1992 the figure was at thirty-eight percent. The percentage of the rushed has grown, and has grown faster and faster.

One consumer desire overtakes another. Consumerism's drug-like hallucinations of happiness rely on the fact that once needs are met, desires must be aggrandized. Faster and faster rates of acquisition of unnecessary products and their fast disposal, feeds, first, manufacturing industry, then landfill sites. Bulimia is indeed a disease of today; consumer society speedily scarfs food beyond need, reaches for the laxatives and speedily excretes. Speed diets are still popular despite being notoriously unsuccess-ful. *Lose your tummy in thirty days on a crash diet.* (Speed and crashes, of course, go together well.) In other ways the body's time is speeded up—two immensely influential childbirth experts in Dublin write, approvingly, that "Prolonged labour, in this hospi-tal, was defined as 36 hours in 1963, reduced to 24 hours in 1968 and, finally, to 12 hours in 1972."

Artists are rapidly overtaking their predecessors in the speed of their production. While the old masters could take years over one work, in the 1960s Andy Warhol was asserting that any painting which takes longer than five minutes to make is a bad painting, and in 1995 Damien Hirst was "creating" art in 90 sec-onds—signing cigarette butts. Fast food is getting faster: five McDonald's restaurants in California have a system which allows motorists to zip through drive-thrus on toll roads and be billed later. The system, boasts McDonald's, cuts fifteen seconds off the normal 131-second wait. Also in the U.S., "Speedpasses" are offered at gas stations, so customers' credit cards are charged automatically. There is a "Speedcook" oven so cookies take only four and a half minutes to bake. For those for whom the sun is too slow, or too annoyingly seasonal, you can buy an "Endless Summer" colorizing lotion which gives you a tan in thirty minutes, any time of year.

In today's socially competitive—overtaking—world, even clothes are at it. From the 1930s the language of speed permeated advertisements for both cars and clothes; one advertiser spoke of garments "which actually streamline the figure." There are quick clothes and slow clothes. Quick ones have sharp lines; most suits and all uniforms are fast, creases ironed to streamline the wearer like a car, company or an economy, to overtake competitors. Slow clothes are for the uncompetitive; baggy jumpers and wafty, billowy clothes, hippy-go-slow clothes on whoa Goa-time. As Western men set the pace of this competitiveness across the world, so it is the Western male's quick sharp suit which dominates the dress-codes of those in the overtaking lanes of the world.

The *Oxford English Dictionary* gives various old meanings for *speed* ranging across "abundance," "good fortune," "assistance" and "help." Speed does, of course, represent the rate of happening; slow or fast. In common usage today, though, it almost exclusively refers to quickness. Speed also once meant success, and today success is indeed defined by speed. To come first in today's society, those in the overtaking lane need fast drugs not slow ones; take caffeine, cocaine, Prozac or speed, which will all pace you up to a crueler competitiveness. Prozac, the *überdrug* of modernity, often makes its users ruthlessly ambitious, speeding past those on the uncompetitive "grass" verges. Prozac is taken for depression and the bereaved are sometimes prescribed it to hurry the stages of grief. It is as if emotional depression—like its financial counterpart—is so unacceptable because it "slows" down a person or an economy. When Gorbachev began the Westernization of Russia, he promulgated three main ideas, two of which—*perestroika* (reconstruction) and *glasnost* (openness)—became famous. The third was *uskorenie*—acceleration. Under Gorbachev, the preceding reign of Brezhnev was officially known as "the era of stagnation."

In such a competitive world, speed is an index to status (in the words of Fats Domino "I'm gonna be a wheel someday"). A first-class stamp betokens more respect than a second-class, while some now sneer at the very idea of "snail mail." The poor travel more slowly; their time is considered less valuable. They are overtaken by the rich and powerful, who are not to be kept waiting; for them the fastest cars, high speed trains and plane shuttles. *Oh, what transports of élites.* "Tell me how fast you go and I'll tell you who you are," as Ivan Illich said. This speed is highly political; one person's speed is paid for by others, so car-centric traffic systems mean pedestrians wait to walk and cyclists are cut up by cars. "Beyond a critical speed," says Illich, "no one can save time without forcing another to lose it." In the U.K., the justification of road-building has been the value of time in the carscape—the car-driver's time is priced at £15 per hour, while the time in the landscape which these roads destroy is given no such value.

The subtlety of place variation is lost at speed; you could be anywhere if you're on a motorway. This is part of speed's deception, appealing to an appetite for change, it offers the opposite—monotony. Fast food is always, monotonously, the same, and one airport, freeway or *Autobahn* is the same as any other. Speed blurs concepts of near and far, leading to what philosopher of transport John Whitelegg calls a "loss of place particularity"—the homogeneity of tourist spots. Slowness, by contrast, the length of time taken to reach a place, operates a "time penalty, protecting place distinctiveness and culture."

Fast travel is a kind of visual consumerism, offering constant replacements of one view with ensuing, newly identical, views.

Travel replicates the model of consumer desires; once first wishes are met, desires must be augmented. The consumer must never be quite satisfied, never be happy. (The name of the car which won the land-speed record in April 1899 was *La Jamais Contente*, The Never Happy.) In 1989, Cesare Marchetti, a Venetian physicist, plotted what is called Marchetti's Constant, which argues that from Neolithic times to medieval and to the modern age, the time spent traveling by people each day has remained at a fairly constant one and a half hours; and though this time stays the same, yet the distance traveled has expanded dramatically, as a result of ever-increasing desires for new journeys and further distances, resulting in greater pollution, more accidents and more social costs.

As Whitelegg says: "People consume the benefit of speed by spending it on distance." Transport studies show that time saved in one journey is used to make additional journeys not previously considered. Mainly in cars. But as Whitelegg points out: "The congestion costs which motorists impose on others are not borne by car drivers." According to Mayer Hillman, Senior Fellow Emeritus at Britain's Policy Studies Institute, increased public transport investment isn't the answer. "Emphasis should be put on walking and cycling. And if that leads to more limited travel," he goes on impishly, "so—fine."

The pollution caused for the sake of speed is also not paid for by the driver. According to the Hillman imp, there is a way to get car drivers to take a taste of their own emissions: "Car manufacturers should be required to design vehicles where the exhaust pipe terminates *within the vehicle.*"

But there is a "green" car. It runs on tap water and toasted tea-cakes and has a built-in gym. It is called a bicycle. "Ah, the bicycle," sighs Whitelegg, "far more sophisticated and useful than anything NASA has ever done." In terms of energy consumption per meter versus body weight, he points out, self-propelled lemmings and passenger aircraft are the least efficient. A body on a bicycle is the most efficient.

"Go Nowhere Fast," says one advertisement, picturing a car in a lush forest, a place of slowness, timelessness and freedom from speed. It exonerates speed, while selling tranquillity. It suggests you can have everything, but is deeply duplicitous; its value depends on the fact that a very small number of people for a very short space of time can have both time and timelessness, can "have" nature *and* a getaway car, selling to those who, by getting there fast, get there *first*. Go Nowhere Fast is for the few a paradox urbane and smug. For the many it is a promise of urban smog.

Car drivers get their benefits—speed and comfort—paid for by other road users in the coin of fear, injury, pollution and congestion. Emotionally, fast drivers get the excitement of speed while their passengers feel the drawbacks of anxiety and powerlessness while pedestrians are killed by speed. An apposite analogy here is with westernized economic structures, where those in the financial driving seat get the rewards of the system, while the dispossessed, without access to the controls, suffer job insecurity, poverty and death. Worldwide, there are "fast castes" and "slow castes" argues Susan George, an expert on the so-called "Third" World debt, showing how the speed with which financial capital moves (more than one trillion dollars a day), benefits the fast but puts the slow into indebted vulnerability. ("We are moving from a world in which the big eat the small to a world in which the fast eat the slow," says Klaus Schwab, head of the World Economic Forum.)

In the name of speed and a "freedom to drive" the actuality is a freedom of *propulsion* but not a freedom of *movement*. Cars can carve up countryside but walkers are not allowed to offer a curious roaming foot to the land, the sensitive and willing foot barely allowed to touch one nudge of earth. Foot-speed gives the roamer a feel of place, while car-speed takes it away. While high speed places must simplify their routes, and offer little complexity or sensory information, walking gives you a great deal of knowledge—of place or of nature or of the people you pass. In walking is wisdom.

Ancient maps are maps of *being*, marking dwelling, Here Be Dragons, Here Be Swamps, maps of place. The Songlines in Australia are maps in stories, a plot for a plot. Maps in everyday use today are very different; they show routes, so modernity's maps are increasingly maps of mobility not of location, geared to speeding movement and antagonistic to being, or dwelling in one place. Highly political, they privilege routes for private, maximum-pollution transport, i.e., cars, rather than giving equal priority to the unpolluting pedestrians and bicycles, or to public transport.

"In India, the cow is supposedly a sacred animal to which motorists must give way. Nowhere in the world is the human being similarly sacred," comments performance poet Mr. Social Control.

Speed-velocity is as hallucinatory as speed-amphetamine—another part of its allure. It is only relative, but its siren-call masquerades as an appeal to an absolute. Absolute instantaneity, the white speed of thought or light, the Zen moment of *inspiring*, breathing in the breath of life. But the danger of speed is in its black opposite, in the instant of *expiring*—the stock market crash, the racing crash, the computer crash, the airline crash, a culture speeding up to its expiration date, the darkness over the event horizon, the moment of death.

Both absolutes share the fascination of the almost unimaginable. The mind can barely hold the understanding for more than a split second, whether it be "seeing" a koan or imagining a black hole. A cultural lust for speed mimics the excitement of inspiration, but in effect it is the morbid excitement of expiry, atrophy accelerated, final and fantastic, the Global Black Monday. As events in the late 90s showed, speed in global financial markets fosters a terrible fragility.

If America is one of the "fastest" places on earth, it is no surprise that in George W. Bush, it found an embodiment of speed. His family's accountant comments: "It seems like he's always wound up and going. He plays a round of golf in two hours," while a friend describes him as "not a sitter-arounder" and he "puts a premium on punctuality" according to the *Washington Post*. "I don't wait well," he wrote in his autobiography, confessing to impatience and abruptness. (Lewis Carroll's White Rabbit winds up in the White House.)

The foreign-exchange market is a speed-conscious place. Meet one dealer at HSBC Midland in London. Hyperalert, fast of speech, eyes darting and breath jerky, he is high on speed, but he takes no drugs; his job drug enough. He follows 70–80 scrolling headline news flashes an hour, monitors a constantly changing aural environment, can turn over £50 million in two to three minutes, and trade with twenty banks in less than sixty seconds.

He drives fast, walks fast, talks fast and eats fast. He says he is "addicted to adrenaline," works "in hypermode" and admits "if you don't enjoy the rushes, you can't do the job." He is a man in love with speed. He describes his faults as speed-related; being easily frustrated by people, short-tempered and intolerant. He hates the speed of human traffic; if he must go down Oxford Street (always crowded with shoppers), he cannot walk but runs at the speed of the buses. (Businesses in Oxford Street intend to fine people £10 for walking too slowly.) His personal calls last, on average, five seconds. Are his friends intimidated by the speed he's going at? "Maybe, yes, but half the time I don't notice. I'm going too fast."

Personal relationships need to develop over time, with time, and speed destroys them, even while providing a substitute. Speed itself is the hallucinatory friend. Speed stimulates, speed stops you feeling bored or lonely. If you can do a hundred miles an hour on the freeway while eating chocolate, who needs sex? Speed bosses the White Rabbit and he bosses Alice, tetchy and intolerant. He has no friends, but he has his watch for company.

Speed harms relationships between people. Western Apaches regard with suspicion any stranger who is quick to launch into friendship; anthropologist Keith H. Basso comments that they "are extremely reluctant to be hurried into friendships—with Anglos or each other" because of their "conviction that the establishment of social relationships is a serious matter that calls for caution, careful judgment, and plenty of time." In Japan, according to the *International Herald Tribune*, companies complain of increasing rudeness associated with speed: fast-food and other service industries with fast employee turnover say that shorter attention spans make it hard for recruits to be taught even basic politeness skills. Far worse are examples of the frustrated speed-need which explodes in road rage. In the nastiest case of its kind in the U.K., Jason Humble, a rally driver with an obsession with speed, was driving into London when he became impatient with the car in front because, he said, it was going too slowly: so he rammed the car over the median into the path of oncoming traffic, killing the car's occupants. It happened, incidentally, beneath a traffic sign which read *Reduce Speed Now.*

Speed can also destroy one's harmonious relationship with oneself; traveling too fast gives you a sort of spiritual jet lag. Bruce Chatwin noted the white explorers in Africa forcing the pace of their African porters who, within sight of their destination, sat and refused to move, waiting, they said, for their spirits to catch up with their bodies. Jeremy Rifkin says we wage "time wars"— the battle between the violent and fast pace of technology and the *lento* tempo which our very bodies require. Roberto, a *tabaquero* of the Ese Eja peoples in Peru says: "You can run, and run at five

· A SIDEWAYS LOOK AT TIME

46

kilometers an hour, but you will see nothing. If you walk, maybe one kilometer in five hours, you will see everything."

Traveling slowly offers more avenues, more choices, more possibilities for meandering or stopping at will. The faster the traveler, the less autonomous they are, the more reliant they must be, for safety, on strict, exterior laws and the directions of systems. Speed fosters passivity. Driving at speed, the individual *is driven* by roads. It is also socially worrisome: individuals accustomed to being told what to do in one arena are more biddable and malleable in other walks—or drives—of life.

The faster you go, the less spontaneous you can be: no pausing, wondering or re-routing. No changing of plans. In the U.K., under the privatized rail service, the cheapest long-haul tickets are for journeys on high-speed trains, booked ahead for a predictable itinerary—benefitting business travel. More spontaneous journeys, made typically on family journeys or friend-visiting, are penalized.

Gentle motion—the relaxed pace of a traditional street, for instance—is hurt by speed. Jean Chesneaux writes: "The street as an art of life is disappearing in favor of traffic arteries. People drive through them on the way to somewhere else." And John Whitelegg says: "English has no positive word for lingering on streets."

His point applies to the social effect of transport, but can be taken further; in English, slowness in general is often treated with pity (a slow learner, retarded), with derision (sluggish) or with suspicion (loitering). To call someone "bovine," slow as a cow, is to express real contempt. Latin yields the wisdom of slowness— *festina lente* (make haste slowly); Italian dignifies it with *largo* or offers the radiant serenity of *dolce fa niente* (literally, sweet doing nothing); while French provides the subversive flirtation of the *flâneur,* the dusky-eyed pauser, stroller and observer. Milan Kundera's novel *Slowness* describes the happy indolence of "the amblers of yesteryear . . . those loafing heroes of folk songs, recalling the Czech proverb: 'They are gazing at God's windows.'" In *As I*   **47**

Walked Out One Midsummer Morning, part of the earthy luxury of Laurie Lee's journey in Spain is the wealth of time the English fiddler found: "There was really no hurry. I was going nowhere . . . Never in my life had I felt so fat with time, so free of the need to be moving or doing." Robert Frost writes: "Everybody should be free to go very slowly . . . what you want, what you're hanging around in the world waiting for, is for something to occur to you."

Diction, in English, reveals an approval of speed in cultural perceptions. Visual perception tells another story. At speed, perspectives are falsified. To speak to the driver, simple little commands of an emasculated language lie in elasticated letters on the road. Spun fast, colors bleed into each other. At speed, foreground is slapped up flat onto foreground. Variations of rhythm and pace are lost, surprise is a hazard, oddness evened out. Subtlety suffers for the sake of speed.

Very like an apple. Chesneaux writes: "The range of cultivated plants has dramatically declined as a result of the race for grossly profitable varieties, for rapid growth. In 1985, 71 per cent of French apple production came from the Golden Delicious variety alone." This is the twentieth century's "Golden Apple," the booby prize in our culture's race against nature, that cliché in the language of fruit, that pappy apology for an apple. Atalanta eat your heart out.

In news media, speed must be a virtue. The London press pressed ahead until recently in the happily apt *Fleet* Street. But the increasing frequency of bulletins on radio, or CNN's rolling permanence, penalizes subtlety, analysis and detail in pride at its speed. The concentration span shortens. Sound-bites bite the hand of ideas that feeds them. These messengers of news, like the original marathon runner, take risks: not so much form drawing level with content—"the medium is the message"—but speed overtaking content—*the marathon is the message.* The runner drops dead, the race run, but the message dies with him. The first victim of speed is truth, and the news *flash* cannot be the truth, the whole truth and nothing but the truth.

All the components of speed today—messengers, fast travel, racing, competitiveness, deceptiveness and a caricature of male sexuality—have a long mythological history. The Greek god of speed is Hermes, messenger of the gods. (His Roman counterpart was Mercury—mercury also known as *quick*silver in English.) God of travelers, of commerce, of athletes and of "fast-talking" deception, to Hermes was attributed the invention of racing, and one of his epithets was "Agonios, who presides over contests"; he was, as it were, the god of competitiveness. The first act of Hermes's life was to steal fifty cattle from Apollo. Hermes the god of speed and the cow, unacknowledged god of slowness, met in an act of mythic theft.

Speed adversely affects language; as the ltle bk of txt abuse called "IH8U" shows too well. "The pen is mightier than the sword—but the keypad is swifter," it advertises itself. The abbreviated words of text messages represent abbreviated emotions, vowel-less and thoughtless, the clichés of mindless juvenile communication, with its tawdry insults of ScmBg, SaDO and Nrd. The phenomenon is MnDNmn.

At speed you can afford no margins of irony, no space for play. Fast language is a faddy fashion victim, buying buzz words, flavors of the month, rapidly becoming overused, worn out and discarded. Verbal speeding shortchanges language. "Be brief." Would that it were as common to see "Be prolix. Be funny. Digress." For the sake of efficient, streamlined transmission, you lose the loose intuited allusive nuances. Speed insists on the cliché, the verbal path well-beaten, the motorway. Language wants to take the scenic route, but freedom to roam is made a trespassory offense and language is taken prisoner by speed, let out only occasionally *en parole*. Puns put speed on pause, slow to read, slow to write. James Joyce, the world's greatest punner, is alleged to have written large parts of *Ulysses* at the rate of two sentences in an eight-hour working day. Were his puns trivial? he was once asked. "No, they are at least quadrivial," he answered, putting the questioner on a four-dimensional pause.

Woody Allen may not have read *Ulysses*, but, he says: "I took a speed reading course and read *War and Peace* in twenty minutes. It involves Russia."

Skim-talking and skim-reading promote skim-thinking. Thoughts summoned at speed are likely to be not the best but simply the first; the habitual response, thoughts automatic as opposed to thoughts idiomatic, reflective or ruminative (the root of which is, of course, "chewing the cud." Respect to that cow).

Signs are that many passengers in the roadster of modernity are suffering motion sickness, and these dizzy dissidents of speed are calling for the vehicle of society to slow down. In France, taking its cue from British road protesters, there is the anti-car newsletter *Moins Vite!*, in Italy, a "Campaign for Slow Food," and in Austria an "Association for the Deceleration of Time": a list to which you could add the time-revolt of French insurrectionaires of May 1968, shedding wristwatches and popularizing the paradoxical one-word slogan—"Quick!" Manu Chao's CD *Esperanza* notes: "*Este CD nació de muchos trabajos, viajes, porros y encuentros. Nació sin prisas . . . (porque las prisas matan)*." (This CD was born of much work, many journeys, spliffs and meetings. It was born without hurry, because speed kills.) Spain's greatest architect, Antoni Gaudí, worked for decades on his enormous project, the building of the Sagrada Familia church. When asked how long it would take to complete, he used to reply simply: "My client is not in a hurry."

There are advocates of slow-knowledge. Slow learners are mocked in contemporary thought but slow-knowledge is arguably more valuable than fast. Fast-knowledge, characterized as technologically-based, applied irrespective of locale, allied to power hierarchies and the pursuit of profit, is competitive and pretends to a Nietzschean amorality. It is single-discipline thinking, for "knowledge," being constantly "superseded" means that only those who are increasingly specialized, streamlined, can keep ahead of their subject. But what society needs today are pluralists, renaissance-thinkers, slower-knowers. Slow-knowledge is shared and multi-disciplinary, unowned, a moral knowledge shaped and molded to a specific cultural context, to a particular ecological locale, a sweet snug fit; niche-knowledge as opposed to Nietzsche-knowledge. (Fast-knowledge assumes that nothing much useful was known before the Enlightenment, and that no indigenous "science" is worthy of the name. Slow-knowledge is not quite so—well—cocky.) Milan Kundera, in *Slowness* comments: "There is a secret bond between slowness and memory, between speed and forgetting."

In Sweden, recently, a man pulled the emergency cord on a train and, when it stopped, distributed leaflets saying "Speed is an unnecessary evil that is destroying our lives and our planet." Other environmentalists point out—in more orthodox ways— that, while sustainable solar energy is wasted, every year the world burns as much fossil fuel as the earth produced in almost a million years. Based on a robber economy, modernity plunders more than nature can renew, pollutes far faster than nature can clean, its pace far outstripping nature's speed. As Rachel Carson noted in *Silent Spring* (1962): "The rapidity of change and the speed with which new situations are created follow the impetuous and heedless pace of man, rather than the deliberate pace of nature." Bill McKibben warns in *The End of Nature*: "The typical projections of global warming over the next century . . . mean the world's climate will be changing at ten to sixty times its natural speed" and adds "Our fretfulness can only grow, for it is the natural world that has always provided our chief images of stability, our necessary antidotes to the 'fast-paced,' 'dynamic,' 'on-the-move' human society."

For the sake of speed, tree-monocultures are planted of eucalyptus, one of the fastest-growing trees, or of speedy pines in mono-rows. The mono principle supports speed. Ford cars: mass-produced, mono-jobbed on assembly lines, made fast, sold fast, driven fast. And drive-thru fast food means the same-every-time McMonoburger. The fast track is the monorail. Monocultures of crops are quicker to farm. "Slow" trees, such as oaks, physically nourish their environment, while fast trees exhaust the earth of water and nutrients.

Speed, from fast trees to fast cars, is an enemy of the earth, as the Italian Futurist Marinetti knew. "Hoorah! No more contact with the vile earth," he said, of the racing-car in 1905. "A roaring auto is more beautiful than the Victory of Samothrace." Marinetti's futurist manifesto boasted of his having been the first to introduce a new aesthetic to the world—"the beauty of speed." In *The New Religion-Morality of Speed* (1916), he wrote: "Slowness is naturally foul," and he promulgated "a new good: speed, and a new evil: slowness." Italian Futurists wanted to straighten out the Danube so that it would flow faster; the natural rivers of time literally made to run for human speed. The cult of speed worships the antagonistic moment, competitiveness taken to the point of violence, as Marinetti said: "We want to exalt aggressive action, the racing foot, the fatal leap, the smack and the punch."

Marinetti worshipped technology, glorified war and supported fascism, this last integral to the whole matrix of ideas he stood for. It is but a quick (goose) step from Marinetti to Mussolini's making the trains run on time; the latter a fact often quoted and seldom considered. There is a nasty, steely connection between speed and fascism. The Nazis took power and they gave the German proletariat transport (the Volkswagen). The Nazis also put money into land-speed record attempts. Henry Ford was awarded a medal by Hitler who admired his anti-Semitic politics, his speed-products and his mono-principle processes. Ford's principles also appealed to Soviet Russia. Paul Virilio in *Speed and Politics*, says: "We could even say that the rise of totalitarianism goes hand-in-hand with the development of the state's hold over the circulation of the masses."

Fascistic colonialism marched along trade routes and roads. The pathless oceans were mapped for colonizers and slave traders. The power of the Roman empire ran along its straight roads. In design terms, today, from *Autobahn* to motorway, any high-speed road, with its huge girders, monolithic walls, cuttings and bridges, a love of concrete *über alles*, is always built according to *which* architectural style? The fascist style. This is true both aesthetically and historically. From 1933, Hitler began his great *Autobahn* construction project. (Propaganda films of this were entitled *Roads of the Future*, *Fast Roads*, and *Roads Make Happiness*.) It was both an engineering achievement and a political statement; the roads, and particularly the bridges, were an expression of the concentration of power, of the subjugation of the individual detail to the whole. Filmmaker Hartmut Bitomsky, who made a special study of this, commented: "all structures would fit together as individual links in a chain spanning the Reich."

Today, the ideology of speed, particularly in its aspect of overtaking competitiveness, is behind the phenomenon of multinationals, today's most fascistic force. Theirs is a politics which brooks no ideological opposition, a totalitarianism whereby one market leader seeks—by competition—to destroy competitors, leading to global domination, demanding uniformity, as speed always does, as fascism always does, and destroying environments or people which get in the way.

Marinetti the Man identified with his racing-car, "ideal shaft crossing over the earth," a relationship which has proceeded apace throughout the century, men and their shafting trouser-arrows of speed. "Men have always been the sex organs of the technological world," wrote Marshall McLuhan. Pumping gas for their hot rods, speed speaks to men in the language of sexual power. Pump up the speed. In Britain, an advertisement for a high-speed train showed the phallus of the train blasting through round membranes of clock-dials, as if it penetrated the virginity time barrier itself. In America, the world's first Pumpkin Supergun, the *Q36 Pumpkin Modulator*, was expected to hurl a pumpkin at the speed of sound in October 1997. "Fire it too hard," commented the *Observer*, "and the pumpkin explodes. Too soft, and it just flops out of the muzzle." Ejaculation is all. And, also in October 1997, pump it up in the Nevada desert as the British *Thrust* Supersonic car penetrates the speed of sound at first 759.333 mph then 766.109 mph. The speed of sound is called, with uncanny macho aptness "Mach" One. One fast car model is the Ferrari *Testarossa*, meaning in Italian "red head," while giving overtones of "testosterone" to English ears.

"MAX POWER: TOP MAD MOTORS, TOP TOTTY, TOP ICE, TOP CRUISES, TOP SPEED, TOP BOOK" is the name of an adolescent book bar none; bright colors, the text pimply with boob-goggling and woman-hating. Intended to be a "Lads'" book about cars and speed, it is a cultural treasure for the way in which it artlessly illustrates society's connections between speed, machismo, competitiveness, cars, power, traffic accidents, sexism and sexual assault. It approves racing at 130 mph on public roads, has drooling descriptions of women in the language of car-specifications, and encourages drunk driving, "doing burnouts on holy ground (synagogues, churches, mosques etc.)". A "joke" section includes one about a female traffic accident victim being sexually assaulted.

What of women? When women involuntarily enter the forcefield of "speed-arenas," they risk being victims of speed-related accidents and sexual threat. (Wham, bam, thank you ma'am.) Entering willingly, their womanhood is savaged and (as in all male-dominated arenas), they must choose one of two extremes, either to be bimbos or *faux*-males. Either masculinized like female athletes and the Thatcher-like women in power, whose power is attained only by aping men, deeply damaging feminism and female self-respect in the process. Or taking the fluff-pot bimbo-choice: cheerleaders at sports stadiums, trolley-dollies on aircraft, whizzy-lizzies at Daytona, waving at the crash-barriers. Women near soldiers must be either Butch or Barbie.

Why soldiers? What has war to do with speed? Everything.

"Speed is the essence of war," according to Sun Tzu, in *The Chinese Art of War*. Speed's vocabulary is violent; high-speed trains are called "bullet" trains. All speed records "break" previous records. The *Thrust* car was reported as "shattering" the previous land-speed record, it "sounded like a gun," as it "smashed" the sound barrier—driven by an ex-Royal Air Force pilot. Speed's mythology is violent—Hermes was the inventor of boxing. The chiefs in speed-arenas of finance and transport have military titles: captains of industry, airline captains, the Governor General of the Bank of England, while the current Punkin Chunkin champion is called Captain Speed.

Paul Virilio writes in *Speed and Politics* that "Western man has appeared superior and dominant, despite inferior demographics, because he appeared *more rapid*. In colonial genocide or ethnocide, he was the *survivor* because he was in fact *super-quick* (*sur-vif*). The French word *vif*, 'lively,' incorporates at least three meanings: swiftness, speed (*vitesse*), likened to *violence*—sudden force, abrupt edge . . . and to life (vie) itself: to be quick means to stay alive . . ."

In English, military orders are executed "at the double." War has its "rapid" troops and the arms "race." ("War has always been a worksite of movement, a speed-factory," says Paul Virilio.) The whole history of war has been a history of the speed of firepower, from a javelin which could be dodged, to the 1866 Prusso-Austrian war where the Prussians could fire seven shots in the time the Austrians took to fire one, with an inevitable outcome. **55**

Recently, a new gun has been invented, the "fastest gun in the West," as one newspaper called it with horribly misplaced jocularity, which fires one and a half million rounds a minute: two hundred and fifty times faster than the fastest conventional gun. But speeded-up weapons only represent half the story, the other is speeded-up diplomacy: the furious, terrible hastening of dialogue, from the First World War onwards. During the 1962 Cuban missile crisis the two superpowers had fifteen minutes' warning of a strike, but then the Russian rockets on Cuba threatened to speed it up to thirty seconds. The speed of warning time in warfare today, counted in minutes and seconds, is something which only children and a scattering of adults are wise enough to really fear.

These generations of children grow up too quickly, under the terror of speed, the four-minute warning. But in many other ways, speed's first casualties are children, who are literally culled in car crashes. In the U.S., the National Center for Health Statistics says that injuries resulting from motor vehicle crashes are the leading cause of death for children one to fourteen years old. It is the same story in Britain. Even without car accidents, children are hurt by speed. Children today can be divided into two classes: not rich versus poor, but "battery children" versus "free-range," according to research commissioned by the Swiss government. It showed that ubiquitous cars meant that most children were cooped up at home, and these battery children were more likely to be obese, more aggressive and to grow up bullies than the—far fewer—"free-range" children who could safely play outside. In another sense, children are speed's victims, hurried through their youth. The *International Herald Tribune* describes how in "today's hectic world," parents want children's free time to be spent on organized activities; sports, homework, music lessons. One result is that the time children spend on chores declined 30 percent between 1981 and 1997. But presumably also reduced is the time children spend on sweet, disorganized spontaneous play. In many countries, education is increasingly compressed, the phenomenon of "Greenhousing," a massive metaphoric G-force, squeezing the childhood out of the child and, in Japan, provoking the highest child-suicide rate in the world. Children's life stages are unnaturally speeded up. In the U.S., accepted medical wisdom states that one percent of girls show signs of breast development and pubic hair by the age of eight. A recent study, however, showed that this was no longer the case, that now one percent had signs of one or both at the age of three.

Our earliest ancestors depended on a need for acceleration, in fight or flight. Our children's survival, by contrast, depends on our judicious, and speedy, use of the brakes. The trouble is that this metaphorical car is being driven by a seventeen-year-old boy, hooked on speed, seeing the world's resources as something to use up before anyone else gets to them. Fatuous adolescence gives us all a spin in its souped-up Ford Capri sooner or later, but modern westernized society is increasingly characterized by unrelenting adolescence.

It is a culture ignorant of the past and viciously refusing to plan for the future, not respecting the old, not cherishing the young. Its exports are adolescent; fast cars, fast food, fast talk, fast bucks, fast war. Fast in everything, puerile and premature, modern westernized cultures could never have produced the *Kama Sutra*, would never pause to consider the point of orgasm maintained for hours. In contrast to the duration of love, and the love of duration (Hermes would have to be the god of premature ejaculation), the West's great love affair is with obsolescence. Jejune in its desire for greed above need, speed above subtlety, it crashes through the gears, cornering too fast, flinging grit in the eyes of the ancient cow, in ancient slowness chewing. In rumination still.

3 · MYTHICAL LIZARDS, BACCHUS'S BINS AND MINING THE PAST

But it's the truth even if it didn't happen.

—KEN KESEY, One Flew Over The Cuckoo's Nest

Once Upon a Time, Princess Diana. The myth, the legend, Diana the fairy-tale princess, *Diana, Her True Story*, Diana, the Grade One Listed Narrative Structure. The Diana-story, like all great stories, was structured in time itself, to make a meaningful pattern out of casual time. Born in full summer, married in full summer, died in full summer. Death-the-storyteller's ending likes chiaroscuro. The Blonde in the Black Mercedes. As her "candle" shines brightest, *phut*, blow it out. At the day's noon, she was on top of the world, more alive than ever, and at its midnight she was dying, underground, in a tunnel. Narrative time was there in all the details; Diana's secret "last" words; those who allegedly "foresaw" the accident, including Diana herself. She took a secret "to the grave"—was she engaged? Was she leaving England forever? Was she converting to Islam? Was she with child?

Tragic narrative needs a sense of time or mistime; "cover her face; mine eyes dazzle: she died young," Ferdinand's reaction to his celebrated dead sister in *The Duchess of Malfi*. Dying young, Diana died dazzling as much as dazzled, the most photographed woman in the world. Marilyn Monroe was, like Diana, aged 36 and the most photographed woman in her world, when she died in 1962, just months after Diana was born. Die young, of course, and stay an icon. They never grow old, forever frozen in time, their candles burning out long before their legends ever will.

The media spoke of Diana's life with a subtext of time—hers

was the wedding of the "century;" a "day" no one would forget; the biggest television event "in history." She was the spirit of "the times;" ours was the "Age" of Diana; she was the star of "a golden age." Her death was a "defining moment"; it ushered in the "week" of Diana; it was the "end of a chapter." Like all myths, she magnified the times, and time was used to magnify her image, even as that image magnified the sales of *Time*, *The Times*, the *New York Times* and the *Times of India*.

Worldwide, two billion people are said to have watched her funeral. Time was stopped in a two-minute silence; the streets were still and trading ceased, at least in Britain. "NO SHOPS. NO FOOTY. NO LOTTO," said the *Sun* newspaper; its front-page usually printed in sunny red, but black for the week of Diana, in a piece of color symbolism—the *Sun's* time stopped, a darkness at noon. Mother Teresa of Calcutta called for two days of prayer time and then, with the exquisite timing of complementary opposites, died herself within days.

Like all subjects of great myths and fables, Diana never chose her meaning; she was the silence at her own storytime. Other voices attached meanings; she was Cinderella, the plain little sister who became beautiful. The then-Archbishop of Canterbury called her wedding "the stuff of which fairy tales are made." She was Sleeping Beauty, kissed by Prince Charles; she was Snow White at her death surrounded by seven dwarfs (the seven paparazzi) on seven motorbikes as they take their pics. Buried on an island she became the Lady of the Lake. Her story took all forms. One day fairy tale, the next iconography: she was "Madonna," "Queen of Heaven," "Saint and Angel," as the cards written in remembrance read. After iconography, the story became Shakespearean. At her Westminster Abbey funeral, her brother Earl Spencer like a Young Pretender arrived on the world stage as if from nowhere, and spoke to the Queen of England, his own godmother, with furious bitterness over Diana's dead body; the applauding audience of the world behind him and a monarch at his mercy beneath his feet. Shakespeare, memorialized in stone in the abbey's vaults below them all, would have died (a second time) for such a plot—a story carved of the very stone of his own dramatic sensibility.

And then again, there were elements of Greek tragedy. The end was bespoken in the very first line of the play: she lived by the image and died by the image. The press—allegedly—gave her her life and took it away. (It spoils the symmetry to deny it; she was actually killed by speed and the car culture, but no matter. The narrative structure, embedded in the human mind, demands that the camera which created her was the camera which killed her. So be it.)

And what of us? The watching world was given something rare—a sense of community in time which myth and ritual have always provided but which modernity usually despises. All radio stations and TV channels, all newspapers and all conversations were for once in sync, with normal programming suspended. Strangers—thinking about the same thing at the same time— spoke to each other; all the ordinary traffic of life stopped.

That week, the ordinary time of usual life met the extraordinary time of myth. Mythic time, says mythologist and historian of religion Mircea Eliade, "interrupts the succession of ordinary time," and the mythic moment is where the profane present meets a sacred eternity. (The ancient Greeks referred to ordinary time as *Chronos*, differentiating it from *Aion*, a sacred, eternal, extraordinary time.)

Traffic and Business define profane, ordinary time today, and for Diana's fabled wedding and "sacred" funeral, traffic and business stopped. It was as if people wanted to know that the usual trafficky, businesslike time of historical consecutiveness is not all that there is. The hysteria which accompanied her death was a hunger for mythmaking—of people fed too much ordinary history-time and not enough mythic, extraordinary, time. Sober commentators bemoaned it—accurately so—saying she wasn't mythic, she was human. Quite right. And quite wrong. Diana-the-person and Diana-the-phenomenon were always entirely different: the historic person was, at a guess, manipulative; rather-dim-but-could-be-bright; both kind and unkind; unhappy, full-feeling, and unpredictable; a silly, funny princess of consumerism, plastic as a charge card. That human-sized historic Diana was not wanted; the mythic phenomenon was what people mourned, and they wanted the myth as they implicitly wanted mythic time.

Diana's brother, Earl Spencer, at Diana's funeral, said: "Of all the ironies about Diana, perhaps the greatest was this: that a girl given the name of the ancient goddess of hunting was herself the most hunted person of the modern age." Diana's very name is mythic, not historic, in vivid contrast to the Royal Family's names, the Princess Margarets and Annes, the Queen Marys and Victorias, Queen Elizabeths and Queen Annes laced with his-toricity, while Diana remains pure myth. People wanted the mythic Diana not the historic royals.

The desire for such a mythic time which stops prosaic time is there in the whole affair of modern stardom, in all modernity's culture heroes and mythic people, from popes to Mandelas, from sports stars and war heroes, princesses and film stars, for all of whom traffic, both real and metaphoric, is stopped, and for all of whom people wave flags. Flags are an international index to culture-heroes; people waved flags for Diana's wedding and half-masted them for her funeral. Football fans wave them for Can-tona, Pelé and Maradona, there are flags on the moon and flags for war heroes and for film stars, flags for all these camouflages of modern mythology.

From Tibet's mountain passes, where prayer flags show the moment the mortal world meets the eternal, to the sports stadi-ums of the world, the flag, perhaps alone of emblems, speaks across ages and across cultures to say to the human mind: "You stand in a mythic moment." In the U.S., "Old Glory flies at half-staff" wrote the *New York Times* in the aftermath of September 11. Flags were planted in the smouldering rubble of the World Trade Center. Kmarts and Wal-Marts and hardware stores sold out of flags as Americans mythologized the day.

The Diana-phenomenon illustrated a hunger for myth and so does the worldwide popularity of cinema, for it performs all the functions of myth. One such function, notes Mircea Eliade, is that myths offer an "opening" into a culture's "Great Time," (when the mythic culture-heroes lived, when the human world was created) and a way of "escaping" the passage of ordinary time. Films, too, nurture that escapist desire and have their "openings" into "great times." For practical and psychological reasons, film screenings begin, usually, at dusk, when stories are traditionally told, the lessening daylight making the legendary past lean nearer the ordinary present. Cinema fantasies are projected, by actual projectors of light, and by projections of the psyche, escaping the finite time of the self for the infinite time of myth. This "Great Time," says Eliade, "is the time that was created and sanctified by the gods at the period of their gesta," that is, their "actions," or "carryings on."

Myths have always depended on a storyteller and today the camera takes that role, joining past and present, memory and moment. When it wishes to hold a mythic moment, as the oral storyteller pauses and repeats a significant detail of the tale, so the camera freezes the frame and repeats the stilled time in the eternal "still" of movie stars.

Myths are often about the beginnings of worlds, paradise being one of the commonest such myths of genesis. Similarly the film *Cinema Paradiso* mythologizes the genesis of cinema, the ancestors of the genre as mythic and as primary as Adam and Eve or the Egyptians' gods of the "First Time." In Aboriginal Australian belief, the stars recall stories of the Sky Heroes of the Dreamtime, the first creation. Many cultures pay homage to the totemic First Creatures in the First Times and so it is that cinema pays homage to its totems from the Classic Age of Film. The totem of the Director's Chair is a chair from that Classic Age, the totemic Camera is never a contemporary film-camera, and the Totemic Typewriter is a Remington.

Myth and cinema both refer to another time, one which is at once true and yet unreal. Supreme illusions both, which spill the present into eternity, their *illusions*—plays of light across the silver screen—are perceived by the story-making mind as *allusions* to another, Silvered time. Silver is the color of Mercury, the mythical messenger and teller of tales, and silver is the color of the myths of the night, playing silver star of eternity to the everyday sky. Diana had that silver quality—some called her "radiant," others "dazzling;" to some she "blazed like a meteor" and she was, said a friend, "mercurial."

As the silver screen must not be touched by muddy fingers likewise the stars of the silver screen are bewitchingly untouchable. Greta Garbo famously wanted to be alone, not to let the muddy finger of ordinary time touch her, for stars, like mythic heroes, are required to remain immortally, iconically The Divine by freezing time in silver. Or freezing it in the silver of death itself, with Diana, Marilyn Monroe, James Dean, Buddy Holly and Princess Grace.

In New Guinea, there is a story about the moon which goes like this. An old woman hid the moon in a pitcher. Some boys noticed it. They crept up stealthily, and opened the pitcher. The moon began to spill out, which scared the boys so they grabbed at it, with grubby hands, struggling to hold it back, but the moon just slid serenely up the sky, though from that day to this, the moon is spotted with the boys' grubby fingerprints. Myths and fairy tales often say: "Look, but don't touch." Touch is the earthiest—most timeful—of senses. Sight is the starriest, and most timeless. Princess Diana's mythic namesake was goddess of the moon, a virgin huntress, worshipped by women; Princess Diana, icon of women, was so famously the virgin bride and, like the goddess, elicited a desire to touch. To her brother, the only blessing of her death was that "Diana is now in a place where no human being can ever *touch* her again."

Back to the moon; untouched by the fingers of little boys and the hands of men, it was part of mythic eternity. From that exact moment of its being touched, by American astronaut Neil Armstrong, its time was tied to man's. And "Armstrong," what a name with which to make such a momentous touch. The moon's timeless integrity (etymologically "integrity" means untouchedness) was lost like the virginity it so symbolizes at the moment it was touched. People called the moon-landing a historic moment, and it was indeed historic; at that moment, the moon ceased to be mythic, for myth is an opposite of history.

The time of myth silvers the imagination of all cultures but the Aboriginal Australian Dreamtime is probably the most magical of all. It is an extraordinary model of time, "extraordinary" by being outside (extra) the ordinary sequence of time which is concrete and actual. Mythic Dreamtime is subtle, ambiguous and diffuse.

The Ancestors—such as the Frilled Lizard Man, Little Wallaby Man and Emu Woman—created the dawning world in the Dreamtime. Emerging from underground, they sang their way across the land, creating its features where they kicked up dust, slept, ate or urinated, then they "went back in" to the land, and became *djang* or energy. Singing those stories is not memory of time-past, but participation in a diffuse, metaphoric, depth of time-present, for the Dreamtime merges past, present and future, as dreams can make past, present and future lap at each other. The

Western mind sees either past or present or future; it only sees the wallaby at the waterhole now, this Wednesday *or* last Wednesday *or* next Wednesday. But the Aboriginal *now* is porous to the Dreamtime Forever, seeing when the totemic Wallaby will be *and* was *and* is, at the totemic Waterhole, in the Dreamtime Wednesday. The Dreamtime or *Alcheringa* is a sacred time, a Great Time, qualitatively different from ordinary time; and while the Dreamtime sustains the present, the present, in turn, sustains the Dreamtime through myth. The Dreamtime is always immanent in the land, which is why mining these lands is a desecration of the Dreamtime.

For the Colombian Pirá-paraná Indians, the ancestral past, similarly, envelops the present. Rather than being cut off from it, the past is an alternative aspect of the present, approached through shamanism and ritual. The Koyukon (of northwestern Alaska) believe in the "Distant Time"—again, not time long ago in history, but time as another dimension of the present. (The timing of the Distant Time stories is also important, for, like most indigenous peoples of North America, they tell their most sacred stories only at night and only during the winter, partly to influence nature's time—to quicken up winter, for instance, so a Koyukon Distant Time story may end with the words: "I thought that winter had just begun, but now I have chewed off part of it.")

Myths and stories across the world have a profound relationship to time. They enchant time, they represent its ambiguity and enigma. As Western folktales begin "Once upon a time," so Aboriginal Australian myths begin with a nod at time, thus: "In the Dreamtime when the earth was young . . . ," "In the time when the dreaming began, a time when there was neither birth nor death," or "In the earliest days when time began." Among the Iraqw of Tanzania, there are many "once upon a time" openings for stories, often beginning *bal geera* which literally means "first days." In Navajo myth and ritual, past, present and future are interchangeable. The Achuar, a tribe of the Jivaro peoples in Ecuador, start their myths thus: "A long time ago, a long, long time ago . . . ," and tell their stories in the imperfect rather than in the cut-off past tense, ending the story in "now." Again, there is no absolute and simple break between now and then. There is a blurred border like a frayed cloud, not a separation of time but the difference of two modalities.

Storyteller Michael Meade begins an Irish myth thus: "Once upon a time, or below a time, in the time that was no time, that is our time, or not, there was in Ireland a king named Conn Mor . . ." Another begins: "There are five directions. East, where the sun rises. North, where there is trouble. South, where you may find a friend. West where all that begins, ends. And the fifth direction; the place where stories come from and where they say Once Upon a Time." (This chapter, incidentally, doesn't make much of a distinction between myths and folktales, because they share the same characteristics with respect to time.) In Mexico, Emiliano Zapata has become an almost mythic culture hero, able to trounce time. Not only are his exploits still sung, but his eventual return is taken for granted by landless peasants in Southern Mexico.

Mythic stories talk time out of mind, charm time and trick time, clogging it or stretching it: fables make time fabulously paradoxical, a stubborn blot on the face of clock-time but true to the time of the psyche, where past, present and future are kaleidoscoped. Time can run counter-clockwise so the youngest child succeeds where the oldest fails, the dawn can be wiser than the dusk and birds can tell the future. Certain periods of time—three days, a year and a day, seven years and a hundred years—are enchanted. In these archetypal tales all over the world, "sensible" time disappears into a wrinkle; a person dips into a fairy hill or disappears for a night with dwarves, but on their return finds that, Rip van Winklish, a hundred ordinary years have passed. The dwarfish figures which inhabit so many tales are themselves like squashed time, at once close to children, but yet grumpy old men, close to the underground past, but able to offer clues to the future; compressed and animated, like cartoons of time.

The Inuit tell tales which begin "long ago, in the future" which is a beautiful expression of mythic time playing trickster to linear, logical conceptions. But folktales frequently play with time. The Arakmbut of Peru have no stories for the beginning of the world; it's as if they always existed. There are only stories for moments of difficulty in the culture of the Arakmbut, i.e., when a sort of history began; only then does the world "begin" but it "begins" with the end of the world. So their stories start "in the end of the world there was . . ." European folktales play with time from "Once upon a time" to "lived happily ever after." "Once" tells of a past eternal, but the eternity it refers to is also a charmed present, just at one remove from "now." Spanish folktales begin: "Once there was . . ." leading the reader into an eternal-present tense, an enchanted present-continuous, a time in the past which still exists. The present, "now and ever after," is the present continuing, life everlasting and even though the individual action is narrated and completed—"that's all folks"—yet life goes on, ever after, back in the now.

The end of stories, the wind-down tenses, the implied-imperfect tenses (giving the narration the atmosphere of slow sunsets and long shadows), return the listener to the normal present of an implied-imperfect world, just like the rolling titles at a film's end return you to your ordinary life. In their endings, Western folktales finish "happily ever after:" so in Dreamtime myths, there is a return to the present, stories may end with the words, "And so it has been ever since" or "And the spirits of the good men still live in the sacred pool" or "and even today the story is repeated around campfires and all know what the Great Rainbow Snake said and did in the Dreamtime." (Spanish folktales can end with a formulaic ending: "They married and lived happily and tossed the bones of their wedding feast at our noses," done perhaps to shock the listeners back to reality, while triggering an immediate connection between the story and real life.)

The human requirement for the cock of life to shake its tail-feathers at death—to transcend mortality—was, mythologists say, the first great impulse towards myth-making. The earliest evidence of myth-thinking concerns burial, and "gardens of immortality" come in spadefuls all across the world. Mythic stories face death, time's most ferociously fearful aspect, and charm the sting out of it, for this reason: the individual tale ends, myths imply, so the individual life story must end in death, but the life of the species lives from ever-before to ever-after. The consolation of life's continuing is most explicit in this end of a Dreamtime myth: "And so death comes, but life always returns." Their transcendence of death is achieved in part by the archetypal nature of the characters of myth and folktales; the totemic Dreamtime figures, the Jack and the Jill of folktales, even the Everyman of morality plays.

Further, the tales themselves become "immortal," living stories retold from generation to generation in an oral culture, from ceilidh to corroboree. Children hate it if the familiar words of a folktale are altered, it disconcerts them because it robs them of the security of eternity; if the story was there before the child, it will be there after the child, is the comforting reasoning.

The reproduction of the story implies the repetition of life itself, repeated through generations, as the generations themselves reproduce. Sex is at the heart of it. Folktales are often deeply sexual, transcending mortality by being rudely, sexually, alive. Cinderella's "slipper" has a sexual interpretation and both the Sleeping Beauty and Snow White are "aroused" by the kiss of a prince. The animal-groom cycle of stories (where an animal becomes a husband, such as the Beast in "Beauty and the Beast" or the Frog Prince) as Bruno Bettelheim shows, represents the altering of attitudes towards the sexual male, so at the beginning, the beast or frog is repulsive to the presexual girl, as the sexuality of the adult is disgusting to the small child, but as she matures, her perceptions of it alter until she can see it as attractive. When she kisses one, it turns into a prince.

In Indian mythology, the *lingam*, phallic symbol of creativity, and the female *yoni* are longingly worshipped. From the seventh to thirteenth century C.E., Tantric philosophy was emerging in India, finding probably its finest sculptural representation in the Konark Sun Temple in Orissa, which was built over twelve years in the mid-thirteenth century. It explodes with sexual sculptures, figures in curling intimacy, dizzy with passion, hot with the heavy perfervid veering sun before monsoon. The whole temple is built as the chariot of the sun god, the sun seen as the originator of time, and the chariot's wheels are a frequent motif. The temple's erotic sculptures, based on the *Kama Sutra*, illustrate the sexuality which makes the wheel of life turn.

In many world myths, sex is at the heart of creation. The Dreamtime is the time of making and creating, overwhelmingly on the fertile side of life. To the Romans, Ceres, goddess of crops, was an emblem of fertility, time sexual, begetting the crops of the world, begetting Persephone. She is not alone. In myth, there is a lot of loosening of circlets, a ludic losing of days in foreplay. Roman gods are gap-toothed goatherds, galvanized into guttural rutting. Sex is everywhere, vulvae wet and wilful; "hymen" nothing like as shy as now, was a god.

Modernity has too little diffuse myth-time, and equally has too little of the sensual, the diffuse erotic aspect of life (even as it has too much pornography). This perhaps accounts for the ravenous hunger for contemporary myths like Diana, Elvis or Madonna—many of which have a sexual component.

. . .

The real natives of mythic time are always children. They do not live in the historical world and are not the contemporaries of adults. It is therefore in the experience of children that any loss of mythic time is saddest, a loss most evident in the fall of folktales and rise of computer games.

Over time, the folktale unfolds its depths of meanings, different to different people and even different to the same person at different times. In contrast, the computer game's repetitive sameness leaves the player stale with its flat numerical rewards, not the fuller emotional rewards of folktales. The folktale speaks uniquely to each child whereas the computer game is identical to every player, its narrative cannot have more than one meaning and remains unchanging over time. Folktales are rich in language, appealing to children of an age to find it magical that one sound can have several meanings; consider a child's swimmy delight in working out which witch is which. Computer games are designed to be totally unambiguous in their language. (They know thirteen but not a baker's dozen and you cannot read between the lines in a computer game.)

Folktales educate children (from the Latin *e-ducere*, to lead out), leading them into the world and into "real time" where children learn that they can gradually—and psychologically—change into an adult. Computer games enclose the child in their own world, airless, stagnant and antisocial, shut them into "synthetic time." Snow White's stepmother is punished for her wickedness by being forced to dance forever in burning red-hot slippers; children are educated into a moral sense. In computer games, people are often punished not for wickedness, but just for being—which can only deaden and dull a child's moral sense.

The folktale curse is that your wishes will come true, its insight is that you must learn responsibility for your desires, that wants can bring complications and that instant gratification is not necessarily wise or satisfying. Computer games—*Battle Zone, Mortal Kombat, Die Hard*—say "kill or be killed" offering a squalid headrush of instant gratification, equating destruction with a sort of febrile success.

Computer games use the binary pattern: wrong or right, stop or go. Folktales use the triad pattern: the third try, the beginning of plurality. The third attempt is the moment of chance or mischance, of opportunity and possibility, it is unpreprogrammed and unscripted and its pattern of "Fail, Fail, Think, Succeed" is a lesson in perseverance and is the time-structure of hope. Computer games' "Succeed or Fail," by habitual motor response, is the time-structure of despair. In folktales, time, sweet unlimited, is on your side. Slowness enhances the telling and no protagonist is punished for taking time; success indeed depends on the sensitivity and even the slowness of their psychological response. The computer game depends simply on the speed of the physical response: time, limited, is your opponent, you must beat the clock.

The tale's narrative is delicately poised, time offering a toehold to truth—it did happen but long, long ago, in a different "mode" of time—a dreamtime—and there need be no antagonism between fiction and truth. Time is weird in computer games, for the unreal event is constantly reiterative. It never happened and yet it keeps happening.

"In a utilitarian age, of all other times, it is a matter of grave importance that fairy tales should be respected," said Charles Dickens. In a fascist age, they can be expropriated: in Nazi Germany, folklore was used to justify both persecution and war. Anti-Semitic sayings of the Middle Ages were used to ignite hatred. Chinese Nationalists and Italian Fascists also used folklore for agitation.

In more recent years, the Disney industry has developed what has been called "fake lore" rather than folklore; one-dimensional, objective, superficial representations of tales whose value lies precisely in the opposite; in their multi-layered, subjective and subtle depths.

Archbishop James Ussher, plump with rhomboid certainty, wrote in the *Annals of the World* (1650): "The World was created on 22 October, 4004 BC at six o'clock in the evening." Dr. John Lightfoot, Expert-Anorak of the Victorian era, declared in 1859 that "man was created on 23 October 4004 BC at nine o'clock in the morning." Newton clung rigidly to Ussher's chronology and sought to use science to prove the Creation myth in *Genesis*.

The past, under Western Christian eyes, must be historic, linear and limited. Stephen Jay Gould declares: "Time's arrow is the primary metaphor of biblical history," while the uniqueness of Christ's incarnation, dividing the whole of time into B.C. and A.D., was crucial to the development of "history" and antithetical to the idea of myth.

Other cultures do not see the past as the Christianized West does. India has its "vessel above time," always full to overflowing, a notion of eternity transcending any temporality. From the Dakota peoples of North America to Aboriginal Australians and Tikopia peoples in Polynesia, mythic time is circular, and is often pictured as the circle of a coiled snake—whether the Norse Midgard serpent; the tail-eating serpent, Ouroboros, in ancient Greece; the Egyptian belief that each person was protected by a "lifetime snake" symbolizing survival after death; or the Rainbow Snake of the Aboriginal Australians which is the source of life. Among the Dani of West Papua the ancestors are considered to be "snake people" able to shed their skins and recover their youth. The Desana people of the Colombian Amazon say that in the beginning of time their ancestors arrived in canoes shaped like huge serpents. They consider the cosmic anaconda as creator of life. Serpent-shaped boats considered as vessels of life at time's beginning are common to mythologies of the Peruvian and Colombian Amazon. Anthropologist Jeremy Narby, in his fascinating *The Cosmic Serpent: DNA and the Origins of Knowledge* comments that creator gods in the form of cosmic serpents recur throughout the myths of not just Amazonia, but Mexico, Australia, Sumer, Egypt, Persia, India, the Pacific, Crete, Greece and Scandinavia.

Time historian G.J. Whitrow comments on the widespread images of snakes: "This symbolism is similar to that found in many ancient cultures, including those of Mesoamerica, in which the serpent represents cycles of endless time, perhaps suggested by the fact that the snake periodically sheds and renews its skin." Whereas mythic cycles mean regeneration and renewal: in the linear historicism of Judaeo-Christianity, time is irreversible, ineluctably deathward bound. Radically opposed to the idea of cyclical time, it is no surprise that followers of the Judaeo-Christian tradition are great snake-haters, maintaining that the destroyer of the first paradise was the serpent in Eden. Narby comments that the Judaeo-Christian story of Genesis contains features common to many creation myths: the serpent, the tree and twin beings, but, for the first time, the serpent is the villain of the piece.

There is more. Narby goes on to draw parallels between cross-cultural myths and Western science. The snake-shaped vital principle is an appropriate description of both DNA and widespread mythic snakes as the creators of life. The double helix, he argues, had in fact symbolized the life principle for thousands of years around the world.

The idea that history is somehow true and therefore superior to myth which is, supposedly, untrue, is highly ideological—and a very self-interested proposition, best suiting those of the Christianized West who invented it. It is an idea which has had its critics, perhaps most succinctly Jean Cocteau: "History is the truth that in the long run becomes a lie, whereas myth is a lie that in the long run becomes the truth."

But mythic time is radically different from historical time and the difference—*pace* Cocteau—does not reside in any distinction between "truth" and "untruth," but in a difference of *quality*. Mythic past is diffuse and ambiguous, its meanings too numinous to mention. Its quality of immanence makes it open to anyone's interpretation, as Diana was, so mythic time has a fundamental democracy, the democracy of dreams, of folktales, of memory; the time of the mind. Anybody's mind. Diana-as-myth was "the *people's* princess," for "the people" did not want the historical—undemocratic—Royals. (And princesses in folktales and myth are utterly democratic for, as every child knows, and every too-politically-correct adult forgets, the royalty is in the *child*; every child is the princess, every child's surroundings her kingdom.) The Western idea of history, by contrast, is undemocratic, not only, famously, written by the victors, but taught by the experts, non-negotiable. In folktales, every child is the expert, as they know in glorious self-righteousness, correcting adults for mis-telling a story. They know how it goes. It goes today like it went yesterday. Exactly like it went yesterday.

(The fundamental democracy of the Diana-myth was affirmed by the hilarious honesty of the public response on the anniversary of her death. The general public was bored of it all. Though Authority, in the shape of newspaper editors, tourism concerns and marketing managers, decreed that the public should re-grieve, the people simply wouldn't oblige.)

Myth, in its etymology, means speech; it—democratically—allies itself with spoken culture and is committed to memory (in Roman mythology, memory is elevated above all muses). The Western idea of history, by contrast, is committed to writing, commits itself to paper and is committed to experts. (It is interesting to see in passing how oral cultures tend to use the clock of "social time," i.e., measuring time by events or by nature, and how the shift from an oral culture to a written culture seems to parallel the shift from that "social time" to "clock-time," i.e., using the abstract, notational description of time.)

One of the distinctions between mythic and historic time could be seen as a difference of gender. Historic time, promulgated by patriarchal Judaeo-Christianity, favored by patriarchies since, uses the linear shape, phallic and male, rather than the elastic and ambiguous time of myth which seems female. The Dreamtime "envelops" the present and Aboriginal Ancestors emerge and return through life-giving moist round places in mother earth. Mythic time is slippery of meaning. Historic time is cut and dried, dry as the papyrus and paper it is written on. History depends on the explicit and visible nature of written language. Myth allies itself with the implicit and invisible. Western historicist cultures and the written tradition use straight line language, avoiding repetition of words or ideas; myth is transmitted through oral speech which uses cyclic language, repeating itself, winding itself back to its beginnings. Similarly, the snake-like mythic "language" of the DNA helix, endlessly repeating itself, twists back to the earliest beginnings of life.

Though myth is distinct from history, the notion that they are implacably opposed is a product of dualist thinking. Anthropologist Andrew Gray describes how the Arakmbut people of Peru view the past:

> The Arakmbut draw no firm line between history and myth; both are termed 'embachapak'—storytelling . . . Both consist of a dialogue with another world (the spirit world and the past respectively) . . . Myth endows history with a powerful position in the collective memory by providing it with a timeless framework of

meaning which itself is constantly changed by reinterpretation according to historical circumstances . . . The relationship between myth and history is not 'either/or' but 'both/and'. The more mythological a story becomes, the more heavily the meaning weighs on the Arakmbut.

Myth is more democratic than history and it is also more democratic with respect to *species*; its vision is ecocratic, giving animals roles and powers, motives and stories, from Aesop to Aboriginal Australians. Everywhere animals, in myth, are considered the old-timers on earth and are often said to be the ancestors of humans, which, of course, proves to be an instinctive understanding of how species, including humans, evolved. The eastern Tukanoan Indians of Colombia say their ancestors are tapirs; the Makuna Indians of southeastern Colombia that theirs were fish people who emerged as anacondas from the Water Doorway to the East of the Ordinary. (And today's largest European tribes believe their ancestors were apes.) The mythic world crawls with animals, teems with coyote-tricksters (the Navajo); with the Badger, the Kingfisher (of the Arikawa); or the Great Turtle and Little Turtle (of the Iroquois); Pegasus-like flying horses; or Ganesh, the Elephant-god associated with time, god of ends and of beginnings. One Dreamtime ancestor is the blue-tongue lizard. Lizards get everywhere in myth, a whole mythology in motion; they are wee dinosaurs, living detail left behind from prehistoric times. Lizards leave their tails behind, as the past leaves its tales behind, as myths leave echoes of their details, a lizardly-like leaving of a tale-tail in your fingers while the rest trails away like a half-heard voice in the night.

Animals are often credited with primal wisdom, perceiving what humans cannot: Native Americans believed that wolves could hear the clouds passing. Time after time (in this time before time) myths say "it was the animals that taught us this." In the folktale "The Three Languages," the so-called "foolish" third son cannot learn from books—that is, from written, historicized and human-centric language—he only takes his education from the three languages of Birdsong, Dogbark and Frogburp. But the moral is that these languages have a wisdom which surpasses solely human thinking.

Across the Northern hemisphere, the bear was once mightily significant to the very passing of time itself, was thought to control the seasons and the destiny of the dead. But this great bear majesty is tweetified today, its animated grandeur reduced to cartoon-time animations; the singing bear Baloo, or Goldilocks and the Three *Reductios ad Absurdum*. In children's time, animals may "live"—though often only in a wrung-out way, as pets or Tom and Jerry cartoons—but meanwhile, "adult" time is writing an all-too-real history of extinction for animals, birds and fish. (Children, who care so much about extinction, are made custodians of animals, as of myths, while the adults, who would render both extinct, have the audacity to call such wise respect "infantilism.")

If modernity is impoverished with respect to mythic time, it is perhaps because facts today *supersede* myth while at the same time, culture *belittles* it. To begin at the belittling.

You could touch down, peckish, at Newark International Airport and fill up your belly with Shiva chocolates, fill up your Mercury car with Shiva Gasoline, check your Midas Muffler, scoot off to New York City, buy a suitcase (a replacement for the one your airline mistakenly re-routed to Dubai, the suitcase's "manifest destiny" according to your airline) from Hercules Luggage, go to the Olympic Deli, spill hummus and retsina on your favorite shirt, try to use Saturn Telecommunication to see whether Aphrodite French Cleaners can clean it, spend some time in Bacchus Greek Taverna reading *Men Are from Mars, Women blah blah,* buy a swimsuit from Athena Swimwear, swim in a Poseidon Pool, visit Chronos Piano Studios (just to hear the piano seller's perfect pitch), go on your way, dazzled outside Krishna Jewelers, and perhaps find yourself gravely leaning on the Atlas fence at the Valhalla Burial Park, Staten Island.

Hop over the Atlantic, take a short urban Odyssey in London. Visit Bacchus's Bins, wine-merchants, and drink until dawn rises rather rosy-nosed as it rose rosy-fingered for Homer of old. Hire a car from the Hercules Car Service, drive past the Olympic Hotel, grabbing a Mars bar on the way, from the Europa Foodshop sandwiched between the Icarus footwear company and Narcissus Ltd., fashion designers. For lunch, try battered cod at the Poseidon Fish Bar, Poseidon's trident spearheading the cod wars, as Trident spearheaded the Cold War. Buy postcards from the Athene postcard shop near the Calypso Voice Agency. Stop at the Valhalla restaurant for an espresso from the Mythos coffee machine, buy a paper from the Krishna, Ganesha or Shiva newsagents. Make a phone call via Mercury Communications. Take a bath and clean it after you with Ajax powder, the myth of old reduced to this, the bathos of bathroom cleaners.

Myth and modernity have a funny relationship. On the one hand, myth comes cuddly to the beckon of *now*, cosily contemporary, so the fury that was Mars has become a chocolate bar and the glory that was Poseidon has become a codpiece to chipshops. (And all the examples given are true.) But, on the other hand, this age is a "supramythic" age.

Modern communications are all but instantaneous, leaving winged Mercury grounded in a lay-by unsuccessfully thumbing a lift in an old, old rain. Bill Gates is not as rich as Croesus, he's far richer (as he demonstrated by having a portrait of himself "made" of money; in a photomosaic, his face was depicted by hundreds of bank notes). The *Pegasus* rocket launcher quite puts the mythic Pegasus out to pasture. As a result of fertility treatment, one

Ms. Chukwu of Houston, an Artemis of Texas, becomes pregnant with eight babies. The mythical figure of fertility, Artemis of Ephesus, is also represented sheltering eight babies.

When Neil Armstrong—he of the mythic, Herculean, name—touched the moon, and left the mythic marker of a flag, it was "a different night from all other nights of the world," said Italian poet Giuseppe Ungaretti. Snaring the moon and stopping the sun are ideas common to myths from Tobago to Timbuktu, but modernity has made it come true—stopping the sun with cloning and snaring the moon with the American flag.

Myth is full of transformings, from Proteus to Cerridwen and Taliesin, but now chemicals transform fish into hermaphrodites (combining in name two mythic beings, Hermes and Aphrodite). Mythology, made redundant by fact, has *gawn fishin'* in a high mandarin sulk, but at the seaside, our rising sea levels could leave all Flood myths high and dry. When biblical myths mustered their meanest might, they promised plagues, deaths and a few thousand locusts, but nowhere in myth is there weather as cruel and frightening as the skies raining acid rain on us. Global warming and the hole in the ozone layer is myth made fact—according to the myth of the Kadimakara at Cullymurra in Australia, there were once monsters who "overconsumed" water and food and destroyed the canopy of trees, until there was one continuous "Great Hole" of blue sky and the monsters then died as a result of the relentless sun. Modernity offers Bhopal, Chernobyl, Hiroshima and Nagasaki, hell's fire on earth and nuclear warfare that will melt your eyes in their sockets, make your skin slip off like loose wet gloves and, if you are wearing a flowery dress, will print the flowers on what's left of your body.

Historians say that an idea of history can be a symptom of cultural insecurity: heritage becomes important when cultures in turmoils of change are losing their traditions. (Take the Romans: a desire for fixity coincided with cultural collapse; the last emperors began calling themselves *aeternitas* just when their civilizations were falling apart, they called themselves eternal just when they weren't.)

It was only after the Renaissance that history was regarded as dead, buried, cut off from the present—and post-Renaissance societies have tried ever harder to dig it up again. For too much change, too much discontinuity from the past makes cultures nervous, increasingly keen to find past foundations to support the present. This is revealed in today's tensely protective attitudes to historical objects. To this end, the laudable Conservationist, diligently repairing the past to a perfection it never knew in its own time. To this end the bony Archeologist, all dust and knuckle, fingering artifacts.

But there is another equally compelling reason, say professors of Heritage Studies. Profit. No one, says the ex-head of the U.K.'s National Trust, protects history for its own sake: in principle it is perfectly proper to profit from the past. Heritage is the biggest growth industry of our time. Ninety-five percent of today's museums were set up after the Second World War, heritage towns are springing up like mushrooms. Nostalgia (that anaesthetized form of memory) is everywhere. Ancestors are—metaphorically— "dug up," antiques are rooted out and souvenirs truffled. It fits, for this is an age of commodification, of the past, as of princesses. And this explains why modernity protects *certain* selected lines of history but not others. Artifact History above all, commodifiable, is preserved. Others, Ritual History, for example, less easily commodified, are nothing like so protected.

Come to Stonehenge at the summer solstice and see a king arrested. King Arthur Pendragon (a living national treasure, one-time caravan dweller, all-time cider-drinker and Excalibur-wielding road protester) believes he was Arthur, the once and future king, in a once and former life and tries annually to reach the Stones to worship at the solstice, keeping Ritual History alive. "We should protect traditions and the root emotions behind them," he barks. "These are important, not the artifacts. The people who built Stonehenge didn't build it for commercial intent but as a place of reverence." (He is annually arrested, in an annual Police Ritual.)

Artifact History is also perhaps valued because it is visible— you can "see" the past, while the audible past is less privileged. The human voice is dateless, so in spoken ritual and in language one can "hear" the past—and can preserve it only by usage. Latin, which is arguably not a dead language at all, but rather in the third stage of its life (Classical, Medieval, New), is unparalleled for authority, *quod erat demonstrandum* (as has been shown), unequalled for the majesty of grief, *lacrimae rerum* (the tears of things) and unsurpassed for the grand shudder, *horresco referens* (I tremble to tell of it). But the present age is unhappily likely to see the last formal protection of spoken Latin dismantled. The oral traditions of storytelling in practice are similarly neglected, while the *written* versions of myths and, of course, the written texts of history are assiduously preserved.

A snapshot from Bhutan, where Lopon Pemala, a cleric who is overseeing the building of a Buddhist monastery, comments:

> Because we believe in reincarnation, our view of history, of the individual, is different. And so is our view of the modern world. Modernity has only one dimension. Museums are institutions of modernity. The past is imprisoned there. In Tibet, the Chinese have made our temples into museums. We do not want that to happen here.

Temples in Bhutan have sometimes burnt to the ground, in fires thought to have begun by butter lamps left burning. Gone are ancient buildings, silk hangings and religious art. But, since here the traditions are "alive" and ritual history never died, its destruction is not so serious. It can be recreated, all of it.

"There is an American penchant to disregard the pertinence of the past," comments Stephen J. Whitfield. He quotes Henry Ford's famous glib remark, "History is bunk," and Huckleberry Finn, archetypal American boy, saying that Moses "had been dead a considerable long time; so then I didn't care no more about him; because I don't take no stock in dead people" and suggests that as far as the past is concerned, Americans live "between amnesia and nostalgia." Whitfield also notes the drop in the number of students majoring in history. "If there is an American amnesia, however, it is undoubtedly connected to the national penchant for zippetty-doo-dah optimism . . . which can transform itself into the politics of hope. Lincoln Steffens's famous line about the Bolshevik Revolution—'I have been over into the future, and it works'—encapsulates the national trait of putting faith in the future rather than revering the past."

In America, news is preferred to history, the present to the past. Novelty is rated; the old is discarded. The new is privileged over the old. The qualities of the adolescent—enthusiasm, dynamism and energy—are sought after, which isn't a bad thing, but the qualities of age are ignored. Which is.

French film-director Jean-Luc Godard believes that France is enduring an American cultural occupation as significant as the German occupation during the Second World War, and one crucial result of this is what he perceives as the loss of the past, the failure of cultural memory. As a corrective, he recommends that television should show "only the past, nothing of the present, not even the weather. They should give the weather from twenty years ago. Tennis matches from twenty years ago, not today. But what's happening today, well, our children will see in twenty years. There's no hurry—twenty years."

The American attitude to the past combines a disregard for the past *per se*, with a desire to commodify history, and to customize it. Umberto Eco comments on the American taste for preserving the past in authentic—life-size—duplications: this Artifact-recreation of the past somehow denies the life of the "real" past by making the present "fake" duplication more prominent, more "real." (The commodification of history can be funny: I once had a query, on a matter of New York's history, to put to the Museum of the City of New York. When I telephoned, the switchboard operator said "I'll put you through to the gift shop: they know everything." They didn't.) Americans lead the world in genealogy-research, the epitome of customized-past, history-as-solipsism. On the Internet, the web's "largest online genealogy search" advertises itself thus: "Find Ancestors and Create a Family." Scientists are also working on so-called immortality chips, so that people will be able to record their lives, allowing them to relive their own and other people's memories. Considering how countries of the West conveniently seem to forget their genocidal histories (more later), it is rum to imagine such care being taken to preserve the "holiday snaps" of an individual life.

My World Atlas lists forty place names in the U.S. beginning with New, from New Albany and New Bedford to New Ulm and New York. Look up Old and the list is a paltry five. (Five nice names, though, including Old Baldy Park, and Old Speck Mountain.) The U.S. is indeed the New World by its own definition: Britain the Old. But for indigenous peoples of the Americas, the true Old World was the life they had before the white colonizers came. Tarzan, in his nineties now, an elder of the Arakmbut in Peru, explains. "*Antiguamente*," he says, anciently, in the old world, life was better. "*Antiguamente, todo era bonito.*" In the old world, everything was good. "Before the contact we lived strong, ate well, went around singing and wearing feathers. We were painted and singing. Yes it was better." Then came the missionaries and colonizers and gave them plagues and sicknesses. Thousands died. (This happened only eleven years before I was born.)

With piebald motives and varying success, what humans have created in the past is often protected. (Heritage, like all historicity, is anthropocentric.) Nature—with the ecocratic ambience of myth as opposed to history—is often less lucky, provoking the question how well would nature be looked after if it had been built by humans, if we could see our own signature on it? Road protesters have cherished Tree History, famously Julia "Butterfly" Hill, living in the canopy of a 200-foot ancient redwood in Stafford County, northern California, to prevent it being cut down. Her tree, "Luna," is a thousand years old.

In the U.K., Prince Philip recently decreed an *annus horribilis* for oak trees in Windsor Great Park; some were to be felled. (Though planted in rigidly straight lines, the older ones had gone in for some gnarly rebellion, and had become "untidy.") This decision was later reversed, but the very attempt was revealing. Windsor Castle was being lovingly and expensively restored after a destructive fire. Of course. It is Artifact History. The castle matters. Whereas the oaks (called, incidentally, "kings of the forest") in the third stage of life, did not matter.

But oaks in that third stage are of most value to other life forms, offering a habitat to bats and owls for instance, and a myriad of microorganisms. Dr. Oliver Rackham, woodland ecologist at Cambridge University, comments that trees develop individuality and distinctiveness after middle age. (Don't we all.) Decay, he says, is not a disease or a defect, but a way of recycling minerals in the wood. The Chinese proverb "Leaf rots to root" expresses the importance of the decaying—the underground—stage of life in nourishing the next, but this, like all "circularity" is not welcome to tidy straight-line thinking.

And King Arthur, in royal tradition, was not amused. "In a

clash between the Oaks and the King, I support the Oaks. It ~~was~~ well out of order for him even to try." He would speak for the trees which can't speak for themselves, though the oaks already felled remain a mute accusation from the past. If oaks could speak, they'd speak in the tone of betrayed majesty. If oaks could speak, they'd speak Latin. If the third stager, Prince Philip, *scampus maximus*, were one of his own oak trees, he would have been cut down.

This difference between attitudes to Artifact History and to Natural History can be applied to modern family relations in the West where dead ancestors are often sought out, in the form of the Artifact Ancestor, the written paper record. This is cherished. Framed. Kept by the family hearth. The Natural Ancestor, the still-living relation (the spittooning great aunt, the completely-barking grandfather) is banished from the family hearth, and left to the barley-sugared etiolation of life in Sunset Homes.

Diverse cultures have always had different attitudes towards the past. In the Holocaust, as well as the infamous and hideous experience of Jews, some 500,000 Gypsies were killed in what they call *Porramous*, the devouring. After the Holocaust, the Jews built cities of memory, memorializing their losses in stone, while Gypsies refuse to remember or discuss it. The Truk islanders of Micronesia in the Pacific Ocean have a culture which means they forget nothing—so they accumulate wrongs; their loyalties and their promises last forever but, equally, so do their grievances and vendettas. The Truk islanders remember everything. The Achuar (of Ecuador) forget everything. They hate and fear the dead. They try to forget their ancestors' names and to lose the memory of the last places where they lived. Stories and dreams are valuable—they live by their dreams, making decisions accordingly, inclining to the future—but memory is feared so that each morning they take a feather and tickle their tonsils, making themselves sick in order to "purge" themselves of the past, in a physical act of forgetting. The Yanesha people of eastern Peru use selective memory when it comes to their culture's past. They remember what makes them different from "others" but refuse to remember any times when those others were more powerful than themselves. Rather than historical facts, what the Yanesha remember are emotions, feelings and moods resulting from their relationships with others, as Fernando Santos-Granero comments. "They obliterate power differences that place them in a subordinate position." They operate a state-of-the-art forgetting in order to de-empower others. They also "sing" their history, one person being the custodian of a certain song. "The transmission and learning of songs establishes a link between past, present and future custodians" while "by transmitting it, the custodian protects the song from oblivion."

One of Europe's special places for preserving history is the British Library, with its experts, authorities, and its authoritative texts. One spring day I visited the historic old Reading Room, with Sure-yani Poroso, of the Leco people. "The West keeps its past written in books," he says. "For us, we don't write our history but we dance it." He is on his feet in a flash to illustrate it. "The dance is a rich literature. We dance how we defeated the Incas and the Spanish and we dance to remember the apocalypse of the rubber era." At our shoulders are books on indigenous peoples of the Americas in a series entitled "Discoveries." The inappropriateness of the term is not lost on Sure-yani. "History is never the story told from the point of view of those annihilated and exterminated. Our history is silenced."

The recording of the past is a political act, and the discourse of history is a form of domination. Perhaps this has never been truer than in the treatment of the history of indigenous peoples, through what the West chooses to remember and what it elects to forget. Sure-yani, and indigenous people all over the Amazon, remembers the genocide where his people were enslaved, raped, tortured, murdered in their thousands, the rubber barons hunting them down with Winchester rifles and hunting dogs, until all Leco people in Peru and many in Bolivia were exterminated. They remember the *barracas*, the concentration camps which very few people ever survived. Native Americans remember the slaughter which reduced their population from approximately five million in 1492 to 237,000 by 1900. In the continent as a whole, writes Jeremy Narby, the most conservative estimate is that 40 million indigenous people died from Alaska to Patagonia—the figure could easily reach 70 million, the worst genocide in human history. They remember.

The Chase Manhattan Bank chooses to forget. Just above us in a small gallery of the British Museum, the Chase Manhattan Bank sponsors an exhibition of Native Americans; dress, artifacts, photos of smiling natives on reserves weaving baskets, happily coinhabiting the continent. History, today, is financed by the victors. The exhibition does make tiny references to isolated "persecutions" but only as if they were sporadic and occasional, rather than widespread and deliberate. It refers to the forced deportations on the Cherokee "Trail of Tears" but without actually mentioning that thousands died. It does note policies of "pacification and assimilation" but without explaining the ethnocidal results of these honeyed terms. It does note that buffalo were exterminated and slaughtered but in so doing it manages to describe the experience of buffaloes more honestly than it describes the experience of Native Americans. The notes mention, as a sad aside, that the Europeans brought diseases to which native peoples had no resistance and that several million died. They note this as if it were all an unfortunate accident, as if blankets were not stuffed with the smallpox virus and as if it were not an intentional slaughter, as if native people were not hunted and murdered, as if they were not victims of a deliberate and vicious genocide.

WITHDRAWN

It is of course a truism to say that history is written by the victors. Alter the stress, though, and the phrase is more telling. History is *written* by the victors. A.R. Radcliffe-Brown, in the mid-twentieth century, promulgated what he considered to be the proper concept of history: the pursuit of institutional origins through the use of written documents. Many indigenous societies ("preliterate" as they were called) were at a stroke considered to be "without history"—simply because it was transmitted in other ways. Spoken, sung or danced history has no validity either intellectually, politically or legally to western eyes. (The Tlingit people of British Columbia and Alaska also danced their history.) "Our dances contain more knowledge than a mountain of books," says Sure-yani. "However, the Catholic church said it was a dance of savages. No one has given any importance to the wisdom we were transmitting, no one, no anthropologist, no missionary." He also describes how the practice of head-shrinking among his people was anything but the grisly, gory phenomenon which Westerners have portrayed. It was a somber, weighty act, done in order to preserve their history, to keep vividly in mind the events of the past when they had been attacked. "A shrunk head was a book of history. We shrank the heads of Spanish people who came seeking El Dorado. The grandfathers would say 'This is the head of Diego Gomez, one of the Spanish conquistadors, and this war drum is made of his skin.' That way our children would remember our history." This way of keeping history was not respected any more than the dances: "The priests came and said we were worshipping fetishes of monkey heads, so the priests burned them."

The clash between two ways of speaking the past, written text versus oral culture, was painfully illustrated in southern Australia, where the Aboriginal Ngarrindjeri women have been campaigning against development of their tribal land (by a development company called—of all things—the *Binalong* company), arguing that it was sacred to them in oral history passed down only by spoken communication and only through women. They were not believed because the state authorities could find no written text and no "experts" to support it. "The white government doesn't believe blackfellas until they see things written down by whitefellas," says Aboriginal historian Doreen Kartinyeri.

Western "rules" of history demand rigid chronology, and treat many types of text as inadmissible. Though academically dominant, a need for strict chronology is an uncommon attitude. Anthropologists' texts describe a myriad of ways of speaking the past. Anna Tsing writes of the history of the world as told by Uma Adang, a Meratus shaman of the Meratus Mountains in Borneo. She delivers history, throwing together many kinds of time, including "folktales, genealogies, and religious tracts, along with imperial and secular chronologies. They fit together awkwardly, challenging one another's assumptions . . . [She] confuses chronology and structure. Her history moves, but it does not go in only one direction. She plays with metaphors of descent and ascent, so that one is never sure whether either metaphor takes one closer to the present."

Hugh Brody in *The Other Side of Eden* writes of Canada's Royal Commission on Aboriginal Peoples in the 1990s, and describes the attempt to encourage the Commission to accept a new form of history which would incorporate the oral history and "stories" and "myths" of First Nations: "a different historical idiom." In particular, this idiom would be used for land rights. (The Royal Commission did not agree to design and fund such a project.) Brody also writes of the longest court case in North American legal history (Delgamuukw v. the Queen) in the 1980s where chiefs of two Native communities of British Columbia challenged the right of the state to jurisdiction over their lands. Their history was oral so they gave oral evidence of their claims; "In the Gitxsan and Wet'suwet'en systems, inheritance of land, and inheritance of the stories that establish rights to the land, are insep-arable." One chief, Mary Johnson, in her eighties at the time, sang their history to the court. "I am asking the witness to sing the song as part of the history, because the song itself invokes the his-tory," said their lawyer. The judge ruled all oral evidence inadmis-sible; these people, he said, lacked all "the badges of civilization . . . had *no written language*, no horses or wheeled vehicles." (My ital-ics.) His ruling was later overturned wholesale.

Eric Wolf describes the political advantage of privileging white, written history and denying the validity of other kinds of history: "The tacit anthropological supposition that [indigenous peoples] are people without history amounts to the erasure of 500 years of confrontation, killing, resurrection, and accommoda-tion." Sure-yani, in the Reading Room, describes how this "blanking out" has made him want to make a representation to Goodyear and Dunlop, the companies which gained their for-tunes by exterminating his peoples.

In the occidental view, the past can be discussed as an ab-straction. All over the world, indigenous peoples see the past as inextricably identified with—and embedded in—the land. "Land is history for us. It is our past, our present and our future," says Sure-yani. We look at one of the books on the Incas, and there is a photograph of an Inca road, coming to an abrupt end. His grandfather had taken him to exactly such a place in Leco terri-tory, where Leco people had defeated the Incas so they came no further. "My grandfather told me to look at this carefully. 'Here is your history,' he said." Leco territory used to be marked on maps of Bolivia, but it was wiped off the map after 1920, "erasing our memory, our history and our land together."

In Peru, the land management organization Instituto del Bien Comun maps land physically and culturally, and various moun-tains in indigenous territory are marked *tiene canción*—it has a song; landscape sings its history. The Quechua concept "Pacha" meanwhile means both "land" and "time" (*tiempo y tierra*), explains Peruvian anthropologist Fernando Santos-Granero. The Arakmbut people in the Peruvian Amazon say, "Without the knowledge of history, the land has no meaning and without the land neither the Arakmbut history nor the culture has any meaning." You find this view around the world. Renato Rosaldo writes of Ilongot head-hunters in the Philippines describing the flight from Japanese troops in 1945: "People were moved to tears as they recited place name after place name—every rock, hill, and stream where they ate, rested or slept." Their history, represented by landscape, could be evoked simply by the power of naming the places in which it occurred. Tribal Filipino Edtami Mansayagan, describing the pain of witnessing the destruction of rivers, valleys, meadows and hillsides of his people's mountain home, says: "these are the living pages of our unwritten history," as Alan Thein Durning reports. In a fascinating article on western Apache landscape, Keith Basso notes that certain land features are a mnemonic peg for myths and stories with a moral point to make. (One story would be related as happening at one particular place, so whenever a person saw that place, they would remember that story, and would immediately feel guilty if they themselves had acted in the way described in the tale.) Time and place are fused so "Apaches view the landscape as a repository of distilled wisdom, a stern but benevolent keeper of tradition."

The indigenous view of the past, then, is different from the Western in representation, in shape, character, significance—and in *vitality*. This is perhaps the most chasmic difference between the Western idea of the past and the indigenous view: the occidental view sees the past as "dead," while the indigenous view sees the past as profoundly "alive." The Australian Aboriginal Ancestors "live" in spite of death: they disappeared, but did not die. They left "indefinite records of themselves" and "images of permanency." They did not "become nothing" but "became the country," as N.D. Munn comments. The past is immanent in the land. "History," says Aboriginal Australian writer Herb Wharton, "comes up from the land." The land is animated with the past, and the past still exists—a different modality of time and one which has a reciprocal relationship with the present. The Dreamtime is opaquely present, inherent and sacred in earth. Sure-yani comments on something similar among his people: "Time is not inert. We live with the past and present together. The past lives in present spiritual values. There is an interrelation between the past and the future, no divorce." Interestingly, many areas rich in myth and indigenous history are shown to be places of high biodiversity; spirited history, life at its liveliest. Both past and present equally vivacious, in a vital land.

The uniqueness of the modern Western sense of the past does not come from any simple distinction between remembering or forgetting, but from a matrix of attitudes; from a distinction between exploitation and integrity, between the factual and the symbolic, between the material and the numinous, between the amoral and the moral, between the profitable and the subtle. Above all, it is in this unique way we have of seeing the past as a dead thing. Cut off. Discontinuous. Dead and For Sale. Other cultures let the past be a continuous surround to the present, a modality of time where life permeates even death itself, life runs its courses underground.

But there's more. The inherently differing notions of the past have direct—and contemporary—political consequences. If the underground past is a source of sacred energy to indigenous people, it is merely a source of literal energy, fuel, to the occidental mind. Mining companies devastate sacred Leco territory in Bolivia. "The land is linked to memory, so you can't take out the gold and minerals. They are part of the body of mother earth. We protest against companies destroying our lands for gold and silver; this is a polemic of memory," says Sure-yani.

Diamonds Are Forever. Made out of the forever-past, they are nature's memory, and to express the fidelity and the beauty of the past, nature says it with diamonds. To express the value of the past, nature has its oils, millennia in the making. As the location of the past is always underground: people "dig up" their family ancestry, looking to "unearth" their "roots;" burials are underground; past memories, too, are "buried;" and with its unerring sense of narrative symbolism, the story of Diana ended with her death underground; so there are worldwide connections between "mining" and the past. Heritage-tourism mines the past for profit. Historians refer to "quarrying" the past, instinctively connecting mining—underground—with digging up history. Freud explicitly compared psychoanalysis and archeology in that both concern the recovery of what is buried.

And to intrude into this underground place is to court trouble. In psychoanalytical terms, it is painful to disturb buried memories. Miners and blacksmiths, using the products of this underground world are, in many societies, marginalized, geographically and socially—they "do everyone's dirty work" and are blamed accordingly. Blacksmiths (with shamans) are the loners of West African Dogon society. In many cultures, blacksmiths live on the outskirts of towns. George Monbiot, in his book *No Man's Land* describes the Nkunono, the ostracized "abominated blacksmith clan" of the Samburu tribe in Kenya. The Nkunono, like many blacksmiths throughout Africa, are said to be polluted, having malevolent supernatural powers. Archeological evidence, comments Monbiot, suggests that European blacksmiths may similarly have been physically and culturally ostracized and equated with evil. Iron Age forges appear to have been built outside other human settlements.

In myths, mining, mineral extraction and metalwork are considered tainted occupations so Hephaestus, the spiteful blacksmith, is lame and mocked. Monbiot draws parallels between

Hephaestus and the northern European Wayland, the "divine but evil smith tutored by the trolls and maimed and imprisoned," and suggests that the medieval vision of Hell represents the smith in his forge. (In pre-Disney folklore, isn't the subconscious suspicion about subterranean work reflected in the subnormal height of the seven dwarves, miners all? And in the Diana-myth, the seven dwarves of the seven paparazzi were called what? Goldminers.)

Today, in this supramythic age, where myths become reality, so these age-old suspicions are amply borne out in fact. Miners mine the—literal—underground past for profit. In Botswana, some of the last of the San tribes may be evicted from the desert lands theirs since time immemorial, to make way for what? For luxury tourism and diamond mining. Across the world, the first and often devastating contact made with an indigenous people has often been for the sake of gold and silver mining. In 1864, thousands of Navajo died, on what is now known as the "Long March," when U.S. government agents forced them to relocate to Fort Sumner, New Mexico, so that miners could search for gold and other precious metals on their lands. The Navajo, who returned to Big Mountain four years after the "Long March," were again evicted from their lands in 1974 when Peabody Coal Company arrived to dig for coal. As Peabody strip-mines in the Black Mesa, and as age-old hostility between the Navajo and the Hopi continues, Native American Daniel Zapata concentrates on the land itself and the sacred past it represents: "The prophecy rock of the Hopi people is down below, right on a vein they want to seam. How would British people feel if someone came and dug up King Henry the Eighth's tomb? Why is it always us?" And the corporations' answer? A merciless manifestation of "mined" over matter—we don't mind and you don't matter. For Roberta Blackgoat, an elfish elder of the Navajo, the past is embodied in the land's song, sung by the ancestors. The coal company is "digging the Mother Earth's liver. The sacred song is going and they're digging, digging, digging . . . My great-great-great-ancestors have been buried here . . . Their prayer is still here, their holy song is still here."

The Mirrar Aborigines of Australia's Northern Territory are fighting the development of the Jabiluka uranium mine, controlled by British mining company Rio Tinto. A government quango notes that there is no effective cultural mapping of sacred sites in this area, apparently due to reluctance on the part of the Mirrar people. Arguably, this is because, to a dualist cartographer in the Western tradition, to name one site sacred is to name another profane . . . and therefore suitable for "development."

Uranium was also quarried in Noonkanbah, Western Australia, numinous site of the Aboriginal Goanna Dreaming (home to the mythic original monitor lizard). The uranium, left where it was, was a source of life to the people of the land; it energized the earth, they said, with *djang*—incipient power, the same power which the Dreamtime heroes poured into the land—and taking it out leaves the land lifeless. (A double lifelessness, for the uranium is used to fuel the death industries of nuclear power.) This *djang* stands for both sacred past and a force of life, a belief found among indigenous people everywhere.

The infamous copper mine at Freeport, in West Papua (Irian Jaya), is on a site considered sacred to indigenous people, the residing place of dead ancestors. The mine is "protected" from indigenous owners by the Indonesian military who have murdered many local indigenous people who oppose it.

In southern Montana, permission has been given for a test well in "Weatherman Draw"—a site sacred to Native Americans in their oral histories: a place for sweat ceremonies and "vision quests." Howard Boggess, local historian and member of the Crow tribe, last Native American "owners" of the site, said: "To us this is a church. In the Native American religion we do not build churches—anytime our feet touch the earth and we can see the sky, we are in our church." Also, there are pictographs on the site which are up to 1,000 years old. The drilling which threatens this history will provide less than a day's worth of current U.S. energy demands.

"We the Guaraní people have our own law of the Mother. The state has environmental and land laws but they seem to count for nothing in the face of the Hydrocarbon Law that is stronger. In January 1999, it was that law that established a petroleum concession in our sacred territory called Jaar where no one is permitted to enter. It is a sacred place," says Bonifacio Barrientos, speaking for the Guaraní in Santa Cruz, Bolivia.

The U'wa people in Colombia consider the "past" of nature in the oil reserves underground as the "blood of the earth" and say that their land is "alive" with oil, coursing like blood in the veins. It is sacred and must be left undisturbed. To take out the oil is to kill the land and themselves, for without the land, they say, they "are not." They say they will commit collective suicide if the plans to extract oil from their territories go ahead. The oil from U'wa lands would, it was thought, sustain global energy demands for a maximum of three months: thousands of years of sacred indigenous past gone to fuel ninety days of the Western present. Occidental epistemology gives rise to occidental exploitation and the U'wa lands have been threatened by the all too aptly named "Occidental" U.S. oil corporation.

Diana's myth, with its dazzling sense of the apposite, included her campaign against mines of another sort—landmines, where the life-giving earth is used as a weapon of death. As Dianabilia becomes a heritage industry in its own right, those who profit from "mining" it are reviled, even while their quarried goods are bought—which precisely reflects modernity's schizophrenic attitudes to miners, we wanted the photos, but hated the goldmining paparazzi. We want the coal and blame the miners with a deep cultural guilt. We want the oil for which the U'wa may die. Conveniently, we can blame those who mine it for us.

It's mine, say the miners—on our behalf. Modernity believes that the underground past is a resource to be exploited, in mining, as in heritage-tourism. The past, being dead, doesn't matter, for its own sake, least of all nature's past, be it old oaks or Goanna Dreamings. To Daniel Zapata, the past is full of life. "From the ancestral past comes life in every sense. When I'm here, I'm flanked by all my ancestors. We're all here, looking at you." An ancient Chinese poem offers a similar sense of the past as alive;

> *The ancestors are listening*
> *Being our guests as they are.*

This is true for two tribes of West Papua: for the Dani, who lay out food and cigarettes for the ancestors in their places in the circle of the living; and for the Asmat, who live surrounded by the skulls of their ancestors who take part in their ceremonies.

Buddhists of Bhutan, rebuilding their temples, have living traditions. The past is not dead. Indigenous Australians of Noonkanbah expressed the living-past thus: "The sacred goannas have been living there under the ground since the Dreamtime. If the drill goes down and kills the goanna it will kill the spirits of the dead." To them, death, if undisturbed, is not dead, because out of death comes life again.

"The past is dead" says the occidental view. "Long live the past" say indigenous people with real grief as they watch Western companies harrowing a past they do not understand, in lands they do not own, killing the living past and hastening the present deaths of peoples wiser—and kinder—than they.

4 · BOTTOMS UP! MISCHIEF NIGHTS AND MILLENNIUM DAYS

People must not do things for fun. We are not here for fun. There is no reference to fun in any Act of Parliament.

—A.P. HERBERT

Feasts of Fools, Mischief Nights and Bonfire Nights once chirped up the whole cycle of the year, wherever in the world you were. Fire festivals, often corresponding to May Day, Midsummer and Halloween, were very widely used to promote the growth of crops and the health of livestock. From Brazilian carnivals to ancient Festivals of Swings in parts of Asia, using see-saws to encourage crops to grow high, from Mardi Gras (Fat Tuesday) to Hopi festivals, serious fun has been a part of the annual round for thousands of years.

Samuel Kinser, writing about carnival American-style, talks of the Mardi Gras at New Orleans and Mobile as a:

Spectacular participation: people paint, stripe, or strip themselves . . . body-paint, spangles, feathers, half-masks, bells and shells, shreds of cloth, ribbons . . . a lovely lady frog plops along in yellow tights, her web-feet made of yellow plastic kitchen gloves.

There are jokers in pointy hats, wildmen, transvestites and an all-prevalent sexual hum. Even the police play; he describes the New Orleans police at the 1987 carnival wearing masks of pig snouts. (Metaphor becomes reality and insult becomes fact: "Look out! Here come the pigs.") A float like a giant would-be dinosaur, called Bacchusaurus, rolls into view, followed by another gigantic figure of Bacchus with a huge horn of plenty, stuffed with grapes.

At a recent New Orleans carnival, one of the oldest carnivaliers remembered seeing the King Zulu figure when he was young: "That king, with moss on his head, then horns like a cow, and a body like a rabbit, and as you went down his body, it would change animals. Man, that was a terrible-looking sight." Another Zulu float-rider carries a long stick with a sign just saying LOUDER.

In Japan, the traditional New Year festival of Namahage was much celebrated in villages, particularly in the rural northeast. The Namahage were young men who, disguised in masks and straw coats and roaring like animals, would search for new brides and small children, pinching them when they were caught. Considered playful, not frightening, they were given *sake* to drink and money or rice cakes, and would visit all the houses in turn, getting progressively drunker.

There is a saying in the German Rhineland that "Whoever is not foolish at Carnival is foolish for the rest of the year." Wine flows in the Rhineland, beer in Bavaria, everywhere it is a time for the mockery of authority, for false noses and fancy dress. In rural areas of southwest Germany, there are strong links between carnival celebrations and pre-Christian festivals. In Rottweil, there is a befeathered fool called Feather Johnny and be-belled Bell Fools. The Schuddig Fools wear fringed clothes, wooden masks and snail shells and they bop anyone they meet on the head with an inflated hog's bladder.

A calendar vivid with carnivals varies the year's course and patterns the social experience of time; all human societies have some form of *off*-time, of carnival or festival, for without festive rhythms, time is too sensible, too well-behaved, too regular and too clockworked. The spirit of carnival is quite the reverse; time-mischievous, time-misbehaved, insensible with inebriation, transvestites galore, festival-time goes pink bananas, stoned, ruttish and—often—vulgar.

In Britain, there were once hundreds of carnivals: blessing-of-the-mead days; hare-pie-scrambling days and cakes-and-ale ceremonies; there were hobby horse days and horn dance days, with their pagan hunting associations and symbolic suggestions of fertility rites; there were well-dressing days, cock-squoiling days (or "throwing-at-cocks"); there were doling days and days for "beating the bounds" of the parish; wassailing the apple trees and playing duck-apple at Halloween; burning the clavie (tar barrel) at New Year, or "Hallooing Largesse" (where, in East Anglia, the Lord of the Harvest traditionally led a troupe of people to sere-nade householders, seeking money), all colored the course of the year. Some of these are pre-Christian, some medieval or later. Many of them have survived in some form—often as "just" a children's game.

At Somerset's Punkie Night, at the end of October, children made punkies (lanterns) out of mangel-wurzels (a large kind of beet) and went knocking on people's doors for money or candles. This was one of the many ancient mischief nights of the year, when children played up gleefully, changing shop signs or taking gates off hinges.

Give us a candle, give us a light,
If you don't you'll get a fright

is the children's refrain; an ancient threat this, playing a trick if you are not treated. Guisers (children disguising themselves at Halloween) in Scotland sang:

If ye dinnae let us in
We will bash yer windies in.

Whuppity Scoorie in Lanark is a festival, believed to have survived from pagan times, during which—in keeping with the New Orleans carnivalier's LOUDER—as much noise as possible was made to scare off evil spirits and protect crops; latterly it is acted out by children who, started by a peal of bells, swing paper balls at each other and scramble for pennies. Up-Helly-Aa is a Shetland Isles festival, dating back to Viking times, when a thirty-foot model of a Viking ship, complete with banners, shields and a bow of a dragon's head, is taken down to the sea by torchlight, then the torches are flung in and it blazes across the water, representing the dead heroes sent to Valhalla in a burning ship. Garland Day at Abbotsbury in Dorset is a ceremony to bless fishing boats at the opening of the mackerel-fishing season, which had strong hints of pagan sacrifice in its thousand-year history, though now it is, like so many other festivals, *just* a children's game. (Not altogether surprising, given that mackerel has been all but wiped out.)

There was bottle-kicking in Hallaton in Leicestershire, crack-nut Sunday in Kingston, Surrey, where once a year churchgoers cracked nuts during services, and in many places there was a Tater-beer Night, a special supper after potato planting. There were harvest festivals with the Kern or Corn Baby, the Hollerin' Pot and other harvest home suppers. There still are cheese-rolling days, that eccentric English custom, ocurring in many places including Brockworth in Gloucestershire. "If you can't hurl yourself down a steep hill after a few drinks chasing cheeses, what's the point of being British? Not even the Black Death stopped our cheese-rolling," said one local. There were (indeed are) egg-rolling days, called pace-egging in Preston, Lancashire, and egg-shackling days in Somerset, when each child would take an egg to school with their name written on it and the eggs would be "shackled" in a sieve, shaken gently until only one egg was left unbroken and the winner got a prize. Now just a children's game, this is believed to have been a fertility ritual.

Customs for festivals in children's calendars very often stop at midday; New Year gifting stops at noon, as do May Day rituals. On April Fool's Day (called Huntigowk Day in Scotland) tricks must never be played after twelve. "It almost seems to children the days of the year change at midday rather than midnight," comment Iona and Peter Opie in *The Lore and Language of Schoolchildren.*

Interestingly, pagan ceremonies begin and end at midday, emphasizing the midnight moment of revelry. A few adult rituals preserve this custom: in some European countries the midwinter festival is celebrated on the night of Christmas Eve rather than the following day. Arguably, acting as our most tenacious custodians of custom, children have actually preserved a pre-Christian festival clock.

Carnivals—and, interchangeably, festivals—are a special instance of the politics of time, for they reverse the norm and overturn usual hierarchies. They are deeply subversive, which is why the powers-that-be are so nervous of them. Consider the U.K.'s (black) Notting Hill Carnival which has been so jumpily and aggressively controlled by (white) police.

For "Y2K" as the year 2000 was being called, by systems analysts and others, politicians prepared celebrations for a millennium-fest which was a highly ideological rhythm of time, though those in charge pretended it was politically neutral. The Christianized West imposed, as ever, its time on the world, and claimed that its time was *the* time. There are many. The year 2000, for instance, was the year 2544 in the Buddhist, 1420 in the Moslem and 5760 in the Jewish calendar. It was not the year 2000. It was *a* year 2000.

Carnival and festival rhythms of social life have long been political, a site of power conflicts, from the ancient battle between Christianity and paganism, to the age of the enclosures, to today. British Puritans, ancient and modern, delight in getting their hands on that filthy little beggar Carnival, washing its smutty mouth out with carbolic soap, making it sit still and behave or be-damned. (Carnival, smirking, would far rather be damned.)

But what is it anyway, this carnival-time? How could you characterize festival-time? Broadly speaking, carnivals have five important attributes. First, they are almost always tied to nature's time. Second, they have an ahistoric quality, not tied to specific events in a recorded past. Third, they transform work-time to play and have a quality of reversal, turning the tables on ordinary social relations, or expected behavior. Fourth, they are characterized by an earthy vulgarity, deeply sexual in their traditions and symbols. And lastly, they emphasize a community of people and a locality of land.

Many festivals chime with the seasons of the agricultural year and of the natural world, the life-death-life cycles of vegetation as, for example, the Obby Oss on May Day at Padstow in Cornwall, where the Oss dances, dies, resurrects and dances again. There are festivals marking the death of winter, or bringing in the summer, there are cyclic (and sacrificial) nature-festivals for the corn spirit wherever corn is grown. It is worldwide; the ancient Chinese, with countless other cultures, believed the crops would not ripen without the observations of seasonal festivals. For the Tewa people of New Mexico (and indeed for many peoples), the spiritual work of festivals is to ensure that nature's cycles continue, so at the summer solstice festival, for instance, a relay race is run east to west like the path of the sun to give it strength (and the racetrack is blessed with eagle wing-tip feathers). Sure-yani Poroso, of the Leco people, describes "Tutili," their ceremony at the spring equinox. "Tutili is a word in our language, Rika Lapacho, for the religion of Cosmovisiones. It means a gratitude to the earth, a re-meeting with the past and it represents wisdom for the future."

Native Americans speak of ceremonies as "life in the sun-dance time." Festivals are not dictated by clock or calendar but by nature's seasons, so there are full moon festivals and ceremonies for the first dew of spring, for the first frost, for the first caterpillar of the year. According to the Korean "Thousand Year Calendar," on Insects' Awakening Day, insects wake and creep out of hibernation into the spring sunshine. That calendar also celebrates Swallows' Day or the Harvest Moon Festival of Ch'usŏk, embedding time within nature; it is celebrated with traditional games including the tortoise game, which helped to promote community spirit, according to commentators. The Kayapo (or Mebêngôkre) people of Brazil celebrate seasonal ceremonies, tied to the agricultural, hunting and fishing cycles so there are festivals for the maize and manioc seasons and for the hunting seasons for land turtle, tapir and anteater. In Micmac society, feasts celebrate events (the new year which happens at the new moon when the creek is frozen) but not an abstract numerical anniversary, 01/01/00.

Ordinary time is historic and linearly evolves. Festivals, ahistoric, stop time by almost unchanged rituals, the annual re-enactments of pockets of time which dip way, way back. (Archeologists say that Halloween, the Celtic festival of Samhain, has been celebrated for at least five thousand years.)

Festival-time transforms the social order, reversing the established status quo in terms of class, of gender and of roles. The topsy-turvy whirligig of carnival-time lets the supposedly lower orders bring in their revenges on the hierarchies and squirearchies. So at a modern office Christmas party the boss may become clown, the cleaner cheeking the company director: while in medieval and Tudor times, the directors of carnival were the Lords of Misrule who ensured that buffoonery leveled the usual rulers. (That said, the times of misrule overthrew the norms in a merely symbolic attack on the status quo, and for a strictly limited period, after which the old order re-established itself with perhaps all the more solidity.)

Carnival in German Mittenwald is known as Nonsense Day.

One tradition is an "old women's mill" in the Market Square. Old women are put in at the top, emerging as young women at the bottom. True to carnival's quality of reversal, this symbolically reverses the passage of time itself. For the Huichol Indians of central Mexico at pilgrimage-festivals, ordinary relations are reversed so the old act like children, women and men swap roles and even emotions are reversed; sadness becomes happiness. For Hopi Indians, festivals are times when clowns appear; they are obscene, gluttonous, satirical—a reversal of all ordinary expectations of human behavior. Their visual similarity to European clowns is also sharp; frizzy-hair wigs, pointy double-pronged jester-hats, turned-up booties and stripy bodies.

Festivals throughout the world are characterized by vulgarity. Gloriously obscene, many carnivals have strong sexual components, from fertility rites to the erotic overtones of games of lifting, of chucking, with swings, upsurges and risings of all sorts. The hobby horse, for example, was notoriously sexual; as was the phallic Maypole and the Green Man's Horn.

Festival time, traditionally, binds communities together, knitting them to their land, each area tootling to its own festive tune, accented with dialect voices specific to certain places and describing a "vernacular-time." Thus one area's festival calendar could have been different from the calendar of a neighboring locale. For the Traditional Seminoles in Florida, the five-day Green Corn Dance is the most important festival of the year. Danny Billie, a Traditional Seminole leader, comments: "The Green Corn Dance defines who we are and what we are . . . It is the heart and soul of our way of life." For the Ainu people of Japan, the ceremony in September to welcome the first salmon is a statement of the Ainu's separate identity.

Fierce local loyalties in Spain were articulated through local patron saints: in the famous procession in Seville during Holy Week, "the escort of a Virgin from a poor parish would glare with ferocity at the Virgin from a rich church in a fashionable quarter. The Archbishop of Seville himself remarked that 'these people would be ready to die for their local Virgin, but would burn that of their neighbors at the slightest provocation,'" wrote Hugh Thomas in *The Spanish Civil War*.

Festival-time could further delineate not only the physical geography but the economic geography of an area, protecting rights of access or land-use, particularly—in the past—in such customs as the "beating of the bounds" of a parish or village. These customs are worth looking at in detail for their history accounts for much of what has happened to festival-time today.

The *Gentleman's Magazine* of 1833 reports that at Scopwick in Lincolnshire, in Britain, "perambulations for the purpose of preserving the boundaries of the parish" took place annually and "boys who accompanied the procession were made to stand on their heads in holes, as a method of assisting the memory . . ." In other parishes, boys would be "bumped" or thrown into streams or beds of nettles. At St. Cuthbert Wells, boys were either whipped or had a chance to "scramble for halfpence" at various points on the route. These were rites of commonality, in three senses; they were of the common *people* (the commoners); sited on common *land;* and marking common *time.*

The beating of bounds, or processioning, as Bob Bushaway says in *By Rite: Custom, Ceremony and Community in England 1700–1880*, "provided the community with a mental map of the parish . . . which was the collective memory of the community." These festivals tied a society to its past, its land and its rights to that land. But, as Bushaway shows, these customs disappeared, up and down the country, as a result of one thing: enclosures. At Scopwick, the perambulations "have been discontinued from the period of the inclosure"; at Otmoor in Oxfordshire, enclosure took place in 1815 and rights of commons and common-customs stopped then. This was not without protest: at Otmoor, hostility to enclosures finally erupted in 1830, when men began gathering nightly to pull down fences, blowing horns as they did so and many, true to the serious-carnival spirit of transvestite reversal, dressed themselves as women. In the Welsh Rebecca Riots of 1839 and 1843–1844, men rioting both in agrarian discontent and against toll houses to charge people for roads, dressed as women—done, suggested a contemporary account, so that they might "possess the sexual power of unruly womanhood." It was a leaderless group, or at least leaders-unknown, and the anonymous "Rebecca" wrote pamphlets on their behalf. If arrested, men were thought to have given their names as Rebecca.

Pre-enclosure, other customs concerned with common land, with the rights of gleaning, wood-gathering or access were vigorously upheld. Cheese-rolling ceremonies, for instance, used festival-time to mark such rights; when the access was denied, so was the festival. At Shapwick Marsh at Sturminster Marshall, a "feast of Sillabub" was held. It was a "joint-stock merrymaking," so one person might bring the milk of one cow, another the milk of three while yet another might bring wine. With the 1845 enclosure, this custom disappeared and many other festivals of commons were outlawed.

Before enclosures, festivals were vigorously convivial, as numerous chronicles show; they were off-license times, drunken, licentious and rude, ranging from midsummer ales to apple-tree wassailing, from autumn mead-mowing to May Day's liaisons. And the Victorian middle classes hated it. Just as land was literally fenced off and enclosed, so the spirit of carnival-time was metaphorically enclosed, repressed and fenced in by Victorian morality; no drinking, no bawdiness, no sex. The common— very vulgar—character of festival was increasingly outlawed and fenced off from the commoners and turned over to the land-owning middle classes in the form of the queasy, fluttery remains of insipid Victorian festival, the etiolated relics visible in today's

Morris Dancers. The vulgar was sanitized. The lewd and the loud were disallowed. The acts and the spirit of enclosure tried to suppress the broad, unenclosed, unfettered, unbounded exuberance of the vulgar at large.

"All friends round the Wrekin," goes a Shropshire toast concerning the Wrekin hill, "and may the devil rain pebblestones on the toes of our enemies that we may know the buggers by their limp." At the Wrekin annual festival on the first Sunday in May, says one snooty commentator in 1864, "various scenes of drunkenness and licentiousness were frequently exhibited, its celebration has of late been very properly discouraged by the magistracy, and is going deservedly to decay."

Coupling, cup-spilling, balls and bells, beery farts, lips on horns and all the really filthy remarks that women are probably more proficient at than men; all this, the vulgar side of festival-time was censored by the rising middle classes. For this was vulgar in two senses; it was rude, for sure, and it was also vulgar in a class context, it was of the people, of the common people. It simply would not do. The priggish Daniel Defoe, with his middle-class pretensions, wrote of the Horn Fair at Charlton, in *A Tour Through the Whole Island of Great Britain* (1724–1726), that

> the mob . . . take all kinds of liberties, and the women are especially impudent . . . as if it was a day that justified the giving themselves a lease to all manner of indecency and immodesty without any reproach, or without suffering the censure which such behaviour would deserve at another time . . . I rather recommend it to public justice to be suppressed as a nuisance and offence to all sober people.

Few festivals are more flamboyantly vulgar than May Day or Beltane. One pagan festival which the disapproving church did not—could not—colonize, it kept its raw smell of sexual license and its populist grassroots appeal (which was why it was such a natural choice for the socialist movement). Vicarless and knicker-less, lads and lasses going into the groves to get a tree for the May-pole let rip the bushy, glorious fornications of May, when all nature is at it. (May sex led to June weddings—the commonest month for marriage—while the full moon of June was called the "mead" moon, the *honey*moon.) Beltane was celebrated with huge bonfires, the Lord and the Queen of the May (who, in the Middle Ages was often a man dressed as a woman) and Spring was personified by the Green Man—the Wild Man, or Jack in the Green. Dressed in leaves he carried a huge horn (enough said). The Maypole, the phallic pole planted in *mother* earth was the key symbol of this erotic day.

Then came the Puritans, sniffing the rank sexuality and de-crying the Maypole as "this stinking idol": and in 1644 the Long Parliament banned all Maypoles. They also objected to the social reversal of carnival; to Puritans, an attack on the status quo was almost as disgusting as sex. After the Restoration, England's most famous Maypole was erected in London's Strand in 1661; a stonk-ing hundred and thirty feet high, all streamers and garlands, mak-ing people wild with delight, it stood for over fifty merry years. But Isaac Newton put a stop to it. In 1717, he bought the May-pole to use as a post for a telescope to penetrate the darkness of night. In the nineteenth century, Victorians infantilized May Day, making it a child's festival to emphasize innocence—of all things.

But the festival of Beltane and the whole spirit of carnival is robust. Coming from the earth itself, it erupts, whether puritans and politicians like it or not. In rural areas, you can still find Beltane celebrated, complete with Green Men and sexual misbe-havior, with Maypoles and Fools. Probably the most fantastic con-temporary pagan celebrations of Beltane are "organized," if that is the word, by environmental activists on "direct action" sites; one such, an encampment with treehouses and tipis to block the building of a new road at Fairmile near Exeter, had a huge bonfire, complete with fire-leaping, fire-breathing, bagpipes and drums. Beltane suits protesters' love of partying, hatred of puritanism, earthy pagan sympathy and politics, for the direct action movement's politics are those of the anti-enclosure movement itself.

For May Day 1999 activists produced a Reclaim May Day manifesto, a flier for which announced a party at the Tower of London, describing it as the "Carnival of the Oppressed," a people's holiday. The flier linked the church's suppression of the pagan May Day with the mass strikes on May Day 1886 across America (in support of the eight-hour day): it decried both the effect on people's time of increased working hours and the way May Day celebrations are discouraged.

The mass trespasses of the British The Land Is Ours campaign (a part of the direct action movement) use the spirit of carnival-time (bring pipes, horns and drums, bring fun) to beat the bounds, as large numbers of the people, many wildly dressed and carniva-lesque, seek common access to common land, bringing the political use of carnival-time to play on the land. Protest sites spring up against road building or airport runway building from Glasgow to Twyford Down, and from Newbury to Manchester, wherever common land is enclosed by developers and fenced off. With a use of carnival as implacably serious as it is bent on enjoying it-self, the fences and gates at Manchester Airport's proposed run-way were pulled down in endless nightly, drunken raids, by people blowing horns, by "leaderless men" wearing skirts and some of these pink taffeta-skirted men, in historical reference, an-swered to the name of—*Rebecca*.

Christianity destroyed what pagan festivals it could, trying to co-opt those it could not. When it could neither co-opt nor destroy, it tried to alter the very character of carnival. While carnival-time reverses power structures of political or priestly dominance, the church's festivals exaggerated them. If carnival-time is nature-based, seeking common land and common time, Christianity took nature out of its liturgical proceedings and brought festival indoors, enclosing it inside the church. Christianity changed the shape of time in festivals so that, unlike the ahistoric, periodic and repeated character of festival-time, the cyclic deaths-and-resurrections of vegetation gods and goddesses throughout the world, which re-enact that time again *now*, the church's festivals were not re-enactment but remembrance of an unrepeatable *past* event; Christ's myth of death-and-resurrection was "unique and historic," taking place not now but "in the days of Pontius Pilate."

In the U.S., the history of Thanksgiving is a good example of time-politics. The first day celebrated as a feast-day throughout the New England colonies was the defeat of the Pequots in 1637. Both historical and religious, it cruelly emphasized the gloating of the conquerors. By the 1860s, Thanksgiving was regularly celebrated with stuffed turkey. The turkey, it has been said, "powerfully symbolizes the Indians."

If carnival-time was frankly vulgar, Christianity rejected anything associated with the erotic. The Green Man's leaves, rightly recognized by the church as a symbol of sexual vigor, were then interpreted as signs of the writhing torment in store for the lustful. From the fourth century onwards, the church tried to ban dancing; in Christian miracle plays, Satan was a dancer and in its missionary depredations across the world, Christianity forbade dancing. Festive fun was hammered out to the thinness of Bible-page paper.

Arguably, it is through Shakespeare that the English know their carnival past: the days of cakes and ale, the rural fabric of time in the peasants' tradition, not dainty but earthy, the jovial, pie-eating drunken blowout, with fools and jokers and clowns. Then came the Puritans. Shakespeare hated these urban shopkeepers who persecuted the festive spirit of his countrymen and women, these black and white businessmen who blanked the colorful freedom of the theaters in his life, and would close them after his death. Carnival shouts "LOUDER": Puritanism whispers "Silence." Some thirty years after Shakespeare's death, the Puritans had ballad-singers arrested and organs, lutes, viols and flutes destroyed, forbidding, whenever they could, the merry and noisy dance of carnival-time. Puritans "strongly prohibited" Christmas and declared it a day of penance. From 1644 to 1652, the Puritan Parliament met on Christmas Day. (After the Restoration, Christmas was allowed once more, and Charles II demanded that American Puritans repeal their Christmas-ban. They did—sixteen years later. In the southern states, Christmas flourished: "frolicking" and "free-form mayhem" prevailed.)

Before the sixteenth century, in Britain, the rhythm of various communities depended on specific fairs and markets; the rhythm of the week was not so important. This gradually changed, which was particularly due to the Puritans' censoring of the social calendar, promulgating the rhythm of the week, flattening the swinging seasonality of time and decreeing instead a mechanical routine of six days' work followed by one day's pray. Many of the festivals they banned were nature-based; Puritan time was *urban* time—bevelled, paved, uniform and gray. Historian Keith Thomas says the Puritan week was "an important step towards the social acceptance of the modern notion of time as even in quality as opposed to the primitive sense of time's unevenness and irregularity."

The history of the Sabbath-rhythm was highly political from the very outset; one of the first acts of the early Christian church was to choose Sunday as Sabbath as distinct from the Jewish Saturday-Sabbath. For the same reason, suggests Eviatar Zerubavel, Mohammed chose Friday as the Muslim Sabbath in avoidance of both the Jewish Saturday and the Christian Sunday.

The seven-day rhythm of the week, of course, has no counterpart in nature—the Sabbath rhythm is a "man-made" construct. Taking a sweeping view of nature as "female," it is interesting if unsurprising that the three biggest patriarchal religions, Christianity, Judaism and Islam, have all cleaved to this non-natural and "non-female" Sabbath rhythm.

What of today? How fares festival? Carnival-time is under threat, from the increasingly widespread business opening on festivals to make more money out of time. In Britain, public holidays are defined by money—*bank* holidays. Today, like latter-day Puritans, big businesses flatten out the year's festive cycle into a virtual straight line; fewer and fewer festivals bend the year into arches; the banal sameness of supermarket time tarmacs festivals. Shops open throughout solstices and Easter, May Day is a working day, Halloween's wise recollection of the dead is *just* for children—no business would pause to respect it.

As Japan has sought to modernize during the last seventy years or so, various levels of government have denounced traditional, rural festivals as feudal remnants, half childish, half barbaric. As in Britain, a Japanese-style puritanism has striven to reduce local festivals in a drive for economic success. Like the Christian church, officials in Japan have either denounced rural festivals as backward, or co-opted those they could not destroy, by promoting them as tourist attractions. Few countries have lost so many of their festive traditions with such efficiency. The Namahage festival is barely celebrated today.

Breaking the flat pattern of all-year business-time is the modern "holiday-time," a telling example of social rhythms: in the U.S., time off is a *vacation* as opposed to a holiday; to vacate where you are as opposed to celebrating where you are. People today take nuclear-holidays, one family, a couple, at the most a small group of friends, who go away for a special time-off, a playtime of their own. In one way, these holidays replace festival-time, in being non-work happy-days, but there is a crucial difference. Traditional festivals meant a whole village or community taking time off together, furthering a sense of community. Today's nuclear-holidays split and disperse a sense of community and the common-time of carnival.

Yet one aspect of non-ordinary, non-work time is waxing strong: "pageant time." You see it in the Lord Mayor's Show, the Trooping of the Color, the Queen's State Opening of Parliament. You see it in the Guilds, in Royal Weddings and the Election of Sheriffs. The fact that these traditions are maintained, at such enormous cost, many with full state pomp and lavish attention to detail (at the wedding of Charles and Diana, for example, the royal horses were fed pills for the week before to color their droppings a particularly telegenic color between yellow and beige) while other festivals are unvalued and neglectfully dropped from the calendar of modernity, is very revealing.

Come with me to the Queen's Opening of Parliament. But we're commoners without invites; we can't go inside, the ritual space is fenced off to us. So stand at the threshold, lean against the crowd-control fence, light a smoke and watch. Stately processions of dignitaries pass in heavy uniforms at a slow, stiff gait. Here, a static tableau: there, the sword of state and cap of dignity. Heralds, in tabards, with coats of arms, genealogically tailored, wait. Arrives a princess, All Stand in the Lords. All Rise and quiet descends for Queen Betty in stately grandeur, gliding across the chamber in a dazzling white and silver dressing gown. ("Cut the mockery. Respect your betters," says a policeman at the fence. "But gi's a light, Guy, fag's gone out.") The Royals. The Lords. The Ambassadors. The Judges. The Dignitaries. The Bishops. All Present. (The People? Nah, that's you and me, the commoners. This pompiscopalian rite is enclosed, keeping us out just as enclosures kept the commoners off the common land.)

But this ain't no festival; no one's drunk for starters. This is pageantry, the enemy of carnival-time and festival. Festival wants people's participation; pageantry wants the people's partition. Carnival-time reverses the hierarchy, mocking those in power; pageant-time exaggerates hierarchy and social dominance, servile to genealogy, its coats of arms are all snob-rampant. Pageantry is for grown-ups; carnival-time is increasingly *just* a children's game. (What this actually means is that children, with their great respect for fun and play, are actually the custodians of the richest, most ancient of our traditions.)

Pageant is ceremony organized from the top down, rather than the "bottoms-up" of carnival; pageantry is London-based, with all its political dominance, rather than taking place in what Londoners still dismissively call the provinces. The history of pageantry coincides with the creation of the first towns, the guilds of craftsmen in medieval conurbations, while festival in its thousands-years-past traditions is a rural affair, nature-based. Festivals are ahistoric, pageantry keeps its history alive and the historicist Christian church sticks like glue to pageantry—each reflects the other, hierarchical, male-dominated and anti-erotic.

Carnival means dancing, means lively movement and fast erupting vitality; pageant's movements are slow, lifeless and stiff, its procession stately, its pages and ladies are in-waiting. Nobody dances in pageants. No one waits, on the other hand, at festivals. Pageantry abhors spontaneity and outlaws accident; carnival-time is trippy, quick, buoyant, funny, noisy and spontaneous.

We The Vulgar are removed, for rudeness is the arch-enemy of pageantry. All Rise for Queen Betty, no bawdy bint she, enthroned in a bottomless pomp which precludes saucy remarks. "Well, Up Yer Bum" says one of the traditional May Day dancers called the Betty or Betsy, the man-dressed-as-woman, a transvestite fertility-symbol—Queen Betty's antithesis—while the Fool holding a bladder on a stick, together with the hobby horse, rudely prods the backsides of gentility. All Rise to pump pomp quite out of court.

So how else does festival fare today? The congregation at a British supermarket, part of the Asda chain, in the autumn of 1997 thanked god and Asda "for all the work they do to serve us," as they celebrated Harvest Festival in the supermarket's vegetable section. At Christmas they held another service. The Archbishop of Canterbury who performed that Christmas service spoke recently of seeing "a sign saying *Glory to God in the Highest* except the 'e' had dropped out so it actually said *Glory to God in the High st.* I liked that."

What's going on? It suits the Protestant church's whole history to drag festivals as far away from raw nature's time as possible, allying itself to commerce. So much for god, but what of Asda? The whole point of harvest-time is its specificity; (you cannot have harvest-time in February), but the whole point of supermarket-time is its non-specificity, merging February with September, and May with either, there is no such thing as seasonal produce, as there is no such thing as seasonal weather. Furthermore, just as supermarkets reduce food variety (there were once hundreds of varieties of apples, now Golden Delicious reigns supreme), so the whole mind-set of modernity's supermarket-time reduces the once-wide variety of time by making one festival a monolithic event; Christmas, the Golden Delicious of Festivals, leaving no shelf space for the Tater Night, the Punkie Night or Up-Helly-Aa' or Beltane.

Because of this, Christmas bears a weight no one festival can happily sustain; hence Christmas stress; too long-awaited, too disappointing. Too much like hard work, with all that shopping, the cards, the carols, the tree, the cooking, the cleaning. Too competitive, too, as the very first episode of *The Simpsons* showed: Flanders' house is alight with an elaborate, expensive lighting display and a giant Santa, waving; the Simpson house has a crummy, dilapidated string of lightbulbs about to fuse. "Flanders is a big show-off," growls Homer, *sotto voce.*

Christmas has become too family-oriented also, making claustrophobic "over-enclosed" misery for many with families and possibly even more for people without. Christmas was once not a family-based but community-based festival, with wassailers and carolers and local feasts. The changeover to a domestic festival came with the Victorian heyday which so damaged community festivals and was marked by a famous cover of the *Illustrated London News* showing Queen Victoria, Prince Albert and their children sitting, appropriately, indoors; *enclosed.* After the Revolution, America was notionally left without festivals or national holidays, and Christmas Day was the same as any other. Although immigrant communities celebrated Saints' Days, there was no single, common event. By that point, even the Puritans had dropped their opposition to Christmas (though they still militated against "excess and profane jollity") and what Americans settled for was, as in Britain, a celebration centered indoors, in the home, *enclosed.*

And what of New Year?

The midnight of New Year's Eve is a time after an end but before a beginning, it is a threshold, time's strange pause in the doorway between two years. It's symbolically appropriate that so many Christmas and New Year decorations are for thresholds, the Christmas wreath at the door, the holly and the mistletoe over lintels.

Janus, god of doorways, was the Roman god of New Year: January is named after him. Janus also survives in an unconscious heritage; his two-faced image, looking both forward and backwards, is wholly appropriate and reflected in two New Year customs; singing "Auld Lang Syne" looks back at the past, while New Year resolutions look forward to the future, at this ambiguous moment. Ovid related that Janus was called Chaos when the elements of air, fire, water and earth were a shapeless heap. When they separated, Chaos became Janus, and his two faces represent the confusion of his origin.

The "first-footing" figure, standing in the doorway at New Year's Eve, is by custom a dark stranger, coming in silence. It is a neat personification of future time; the unknowable, inaudible arrival, a figure of elastic ambivalence, who might be your traitor or defender, your hangman, confuser or friend. Riddled with ambiguity, time can bring disaster or opportunity but the only certainty is that it will bring change.

A sense of change is crucial to the idea of New Year, for time then is at a junction, a crossroads, and change is more easily accomplished at these forking points. Janus, of course, was god of crossroads. The New Year resolution takes the calendar's turning point of symbolic change, using it as the location for real change. In many ancient societies, says Mircea Eliade, a New Year represented the annulment of the past year and the possiblity of "purifying" what the past year had soiled. In Tibet, it is customary to wear new clothes on the first day of the new year. At the big New Year, 2000, a sense of grubby, guilty, soiled *fin de siècle*-ism gave rise to a newly articulated yearning for purity, with many social commentators talking of a new morality, a new purity, a new fresh start.

To make a resolution is to "turn over a new leaf" and custom and nature here intertwine; the festival at the turning point of the year encourages both metaphoric new leaves and nature's literal new leaves. In the dead of winter, nature's energy is stored but unreleased and the New Year festival can be understood in part as an ancient memory of sympathetic magic; the energy stored in the champagne bottle is uncorked, the frothy spurt of bubbles releases the seminal energy of spring. Paganism persists, *pentimento*. Mistletoe was said to be sacred to Druids who saw the white berries as the spunk of the sun god. Christianity, with its usual squeamish aversion to sex, purses its lips at mistletoe.

The midwinter festival has always been characterized by release; audacious indignity flaunting itself uncorked. The medieval Lords of Misrule released society from everyday restraints for the twelve days of Christmas. Shakespeare's *Twelfth Night* (twelfth night was the old New Year's Day) trips with the footloose carnival spirit. In ancient Rome, Saturn's statue—associated with Chronos, or regular time—was fettered throughout the year, his hands bound with woollen strips, but at New Year, these bands were untied in a symbolic yearly liberation. Time set free. The Roman New Year was called the Saturnalia, in his honor, and at the statue's release, there was a huge social liberation; Romans were set free from public affairs, law courts and schools, while servants were released from their servitude in a carnival freedom. (Apropos of liberation, the Christmas Liberation Front—it exists—outraged at the commercialization of Christmas, removed a Renault car logo from the top of a public Christmas tree in Manchester, in Britain, and replaced it with a star, to set free the spirit of Christmas.)

A year and a day is a time period understood by the human body itself—in a year and a day, the entire natural cycle is complete and you start again. "A year and a day" was a traditional moment of release—that word again—from several obligations: the binding time of a pagan wedding, the hiring period of laborers at trade fairs, the time during which a murder charge could be brought and, by tradition, the time limit of a curse. There is something humane in an annual limit; *forever* can be cruel and endings can bring a quality of mercy, unrestrained. Some societies, such as the North American Micmac people, hold a feast a year after someone has died and part of the reason for this is to release the mourners, to end the year of grieving.

In a spiritual context, release means the cancelling of old sins and the mercy of forgiveness. Jews are required to seek the forgiveness of others at the Jewish New Year. Catholicism stresses the importance of asking god's forgiveness before the end of the calendar year. In today's psychodynamic liturgy, a New Year resolution offers a release from learned responses and habitual behavior patterns and a chance of self-forgiveness. In all three cases, the seeking of forgiveness is only worthwhile if a new beginning is allowed, if the threshold of a new year becomes a place of release from the past.

But symbolically, the place of such release is also one of risk. A junction is the location of both maximum freedom and maximum anxiety, ambivalent as the crossroad was to the Romans. The crossroad was the place of opportunity, but also the burial place of suicides and criminals, place of both chance and mischance, a place between departure and return, a halt between two roads. Similarly at the New Year midnight, time stands at a turning point, with one year ended, another unbegun. (Janus was, incidentally, god of ends and beginnings, of departures and returns.) At a junction looms crisis. Chinese has represented the word *crisis* as both "danger" and "opportunity" and this is at the heart of the feel of New Year. "Opportunity," incidentally, comes from another threshold word, a doorway word, *portus* in Latin which means both door and harbor. (Guess who is god of harbors? Janus. And the Latin for door is *ianua* from—Janus.)

There is a boomerang relation between beginnings and endings: so a beginning contains an implicit energy which draws its ending around and back into itself. But there is no similar energy in an ending, no inevitability that an end implies a beginning. Arguably, therefore, the year's end provokes an unconscious anxiety over whether time will continue and the annual wake for the death of the old year is an apprehensive wakefulness to watch whether the fragile phoenix of the future will arise from the ashes of the past. It is a gap in time which can make the human psyche awed and lonely. The tradition of holding hands is a clutch at comfort and is also, perhaps, another instance of sympathetic magic, an attempt to encourage the new year to join hands with the old, to take time by the hand and pull it in, through the door-way of Janus.

All of these emotions were apparent, in ever-magnified form, with the millennium, the biggest New Year in history. *Torschluss-panik* is how German, with its reliably brilliant abstract nouns (think *Zeitgeist, Weltschmerz* or *Schadenfreude*), describes the panic as the gates are closing, the gate of a year and, then as never, the gates of the millennium. In 1997, just as 2000 was really begin-ning to seem close, came news of the thirty-nine suicides of a UFO cult called, fittingly for this threshold-*zeitgeist*, Heaven's *Gate*.

In the symbolic world, the threshold is a place of drama and struggle and a place where a toll must be paid. Around the world, hotels and restaurants set huge charges for the millennium night and governments, similarly, spent hundreds of millions of dollars as if they were responding to a deep atavistic need to pay a fee at a symbolic gate.

There had been no date in world-time quite like it. There is a popular belief that the year 1000 C.E. was marked by boiling skies and gaping wonders, by naked hermits putting the wind up overemotional peasants who would follow any toothless prophet dribbling nutmeggy remarks and visionary cabbage into his beard. Professor Richard Landes of Boston University's Center for Mil-lennial Studies says that the evidence remaining today (preachers announcing the end of the world in 1000; pilgrims rushing to Je-rusalem in the hope of watching the Last Days) is just the tip of the iceberg. Hostility to millennial ideas produced "a consensus of silence." However, according to Damian Thompson in *The End of Time* the year 1000 passed without much remark simply because the great majority had absolutely no idea of the date. Medieval historian Michael Staunton notes saliently that millennialism permeated the times; plagues and famines were ever-present. One of the earliest noticed turns-of-the-century was, says Thompson, likely to have been 1600, following the publication of the hugely influential *Magdeburg Centuries* (1559) which for the first time drew widespread attention to the idea of counting history by hundreds of years.

The year 2000–2001, though, predicted Harold Bloom in *Omens of the Millennium*, would "not be a comfortable one in the USA because there are extremist groups among the premillennialists . . . The Aryan Nation and similar fascist apocalyptics could seek to assuage unfulfilled expectations by terrorism." The National Health Service in Britain said it expected a surge of mental illness at that time—people suffering from what they termed "millennium panic." There was a sheer power in the numbers, too, the zeros act as an absolute constraint and such a sense of limited time, no matter how arbitrary, has an ability to provoke anxiety in all but the hardiest.

Damian Thompson repeats Paco Rabanne's wild New-Age list of end-time beliefs: we are living in the dying days of the Indian Kali yuga cycle; we are at the end of the Greek Age of Iron; the Mayan calendar will run out on 12/22/2012, etc., only to pour his half-acidic, half-amused scholarly mockery over them. Other commentators who do not share Thompson's care or discrimination make scary lists of end-times: of Mithraic time ending on or around the year 2000; Cabbalistic time ending around the year 2000; an Egyptian cycle which ends in 2001; Nostradamus's terror-prophecy about 1999 and Druid calendars ending in or around, well, the year 2000. Newton speculated on a Second Coming in the year 2000. A recent book, *Fingerprints of the Gods*, selling an epic four and a half million copies, predicts the world's end in 2012.

So what? The point is not whether your Nostradamus is right or my Druid is wrong, the interesting thing was the anxiety itself, the *Torschlusspanik*, and what it reveals about ourselves, for now as ever (and now as never quite before) our attitude to time is—blissfully unwitting—self-portraiture. Obsessed by big numbers, allying time-measurement to priestly or demi-priestly power, and with a seeping guilt about a number of things from environmental crises to the guilt implicit in Christianity whose festival this millennium is.

The whole millennium-fest was a carnival of modernity, and the notorious "millennium bug" was the unpredictable stranger at the party, wearing the carnival mask—is it jeering or smiling? Janus, after all, is two-faced, untrustworthy, your best friend when he works for you, your worst enemy when he works against you. (Very like a computer.)

As a result of computer programming in the 1960s and 1970s, dates were rendered by two digits (69), not four (1969). This was a shorthand done to save memory space and, therefore, money. (It was also done in the belief that well before the year 2000 the technology would become obsolete.) But, as a result, the year 2000 could be read as *year nought,* coming before 1999 or 1969, provoking wild anxiety. "Apocalypse 2000!" and "Doomsday 2000!" came the headlines, while 9/9/99—a "rubbish date" for computers—was something of a dress-rehearsal.

Another headline: more than a year before the event, the British government was advising people to stockpile food and water. The Canadian Armed Forces announced they would be on standby for civil chaos as a result of power losses. Some said Y2K was a matter of just a few hitches and hiccups. Others that it would be a "nightmare scenario." (My computer was fine. For some reason known only to itself, it decided a few years ago that it was 1956, and 1956 it has remained ever since. Every year.) Using time as ever to reflect humanity, optimists were optimistic in the face of the Y2K bug, pessimists pessimistic. Pooh-poohers pooh-poohed the entire problem, arguing it was hyped up by those who would profit from a panic. Others were taking a head-for-the-hills approach, saying, for example, that any job classification that didn't exist in 1945 wouldn't exist in 2001. Buy a cow.

A report on the Y2K bug was published in November 1998 by the independent agency, the British American Security Information Council in Washington. Suggesting that the risk of a nuclear Doomsday scenario was small, the report said nonetheless that although a catastrophe would not be in the form of an unauthorized launch it could come in the form of a nuclear accident. From other sources, posted on the Internet, there were warnings that the coolers of East European nuclear power plants could cease to function properly as a result of the Y2K bug—with catastrophic results. The BASIC report urged that all warheads be physically removed, or decoupled, from their launch pads so that they could not be fired.

This was a vivid parable of our times and of our idea of time. Modernity, in nuclear warheads above all, shrank time to an urgent jerky stump, with its attitude of "launch on warning," the "split-second reaction time," the "hair-trigger response." Too little time was left between the knowledge of a Y2K problem and its arrival: not enough time was left to ensure absolute nuclear safety. Steps should be taken immediately to "de-alert" nuclear forces, said the BASIC report. "These steps would reduce the alert status and increase by minutes, hours, days or weeks the amount of time required to launch a nuclear attack." How beautiful those minutes, hours, days or weeks sound in this context. For, having shortened and brutalized time so nastily, it is only in time itself that there is any forgiveness, any generosity, reprieve or grace.

On the Internet, the bug was referred to by the acronym TEOTWAWKI—The End Of The World As We Know It—a half-jocular, half-nihilistic expression so apt for this edgy unknowing. Every New Year draws attention to the future's unknowability, the dark stranger at the first footing could be your lover or destroyer. The millennium New Year (personified in the bug) merely magnified that unknowability. But this crisis was (as the Chinese knew) both danger and opportunity. Nuclear warheads decoupled from their launch pads could, it has been argued, provide a safer world than we have known for years, while the bug's very existence exposed the fragility of modern life and increased sensitivity to locality in many ways, from food production to the importance of *communitas*.

Most interesting of all is that it fits so perfectly the time-myths society has constructed, the fear at the limits and thresholds of time, Armageddon at the bible's end, the *Torschlusspanik* at the year's end and TEOTWAWKI at the millennium's end. Modernity divorced time from nature, inventing or constructing its own time instead—in this case elegantly and bitterly so. Through computer engineers, modernity wrote a complete fiction of time, *00*, and then imposed this fiction onto the running of the actual world; so this coded artificial construct is given an unforeseen, all-too-real power over humanity—and becomes a self-fulfilling prophecy. The zero, the nothing, the nought—so important to the history of mathematics—turns round and bites the hand that drew it. O, the original vicious circle.

We make time in our own image. In this instance, we are too shortsighted a culture, too speedy, greedy (bigger-bang-for-the-buck) and far too short-termist. In planning obsolescence, we then suffer the consequences when the technology is, for once, not superseded. We wrote the script of our time in shorthand. Literally. *00*. And gave ourselves short shrift with this shorthand; sold ourselves short.

The relationship between humankind and time is sinuously reflexive. The "built-in bug" was actually in us, that bug of fear (which we have harbored for hundreds of years) that civilization will end. The computer *virus* was the virus in us, the feverish pre-millennialism which had us frenzied with panics see-sawing between the groundless and the well-founded. In a sense, these articulations of doom are carnivalesque, but shockingly, appallingly altered. Carnivals are the celebration of time embedded in the natural world; this carnival of modernity was created by the time of technology. If carnivals champion locality, this Y2K bash was a carnival all-too-global in its reach. Carnivals are ahistoric and speak of cyclic time, repeating their festivities year after year. Y2K was the epitome of linear, historic, one-off, unrepeatable time. Modernity has chosen to count time, rather than cherish it. Counting down like a bomb. Very like a bomb. For, albeit widely discounted by experts, the worst fear people talked of with respect to the Y2K bug was that of a literal nuclear countdown. Countdown. Three, Two, One. Two, One, Off. Too one-off indeed.

Carnivals embody the spirit of light-hearted social reversal. For the Y2K global carnival came predictions of all too serious reversals: violent social disorder. The whole affair had a glittering, ghastly fascination, a merry morbidity, see modernity's grinning skull doing the fastest, weirdest dance of excess, a St. Vitus's dance of death, drunken and deadly right up to the edge of time as we have prescribed it, the dancers half-defiant, half-terrified. Behind Carnival's darkest, cruelest-ever mask, a voice was screaming that fear is part of the process. Be afraid before you can be free again. Feel the terror of a world pause before time lurches on once more.

And what of Christianity, with its special relationship to Y2K and its history of counting time? Give the church its due—some Christian leaders, together with environmentalists and social justice groups (in an umbrella movement called Jubilee 2000) were making good use of the guilt that they find, the guilt that everyday folk leave behind every day of this soiled century, and they called for the year 2000 to be celebrated by an act of beauty and appropriateness. Drawing on the widespread notion of New Year as a time of setting free, they said 2000 should be marked by the biggest imaginable release from modern slavery, the cancellation of the Third World Debt—an act as symbolic as it was overdue. The *jubilee*, by Jewish tradition, is the year of emancipation once every fifty years.

But branches of evangelical Christianity, with neo-medieval ignorance, were rather preoccupied with a crusade of cultural genocide, taking as their cue Matthew 24:14. "And this gospel of the kingdom will be preached thoughout the whole world, as a testimony to all nations; and then the end will come." The end of this world is the beginning of—their—everlasting life so, in their logic, by stepping up their preaching of the gospel across the world they can—and should—hasten the end of the world. (AD2000 is an evangelical initiative to coordinate missionary work using the new millennium as a "deadline"—in two senses of the word.) Missionary activity becomes "particularly unpleasant when they target the most vulnerable societies left . . . to gain control of their minds, hearts and spirits, and souls too. The Summer Institute of Linguistics is a worldwide missionary organization dedicated to this kind of wickedness," writes anthropologist Alan Tormaid Campbell. The Summer Institute of Linguistics is a virulent Christian evangelical group widely accused of cultural genocide through its brutal inroads into some of the most fragile tribal societies in the world, bringing self-doubt, fear and panic on a scale these societies have never known. Believing the end of the world will come when the gospel has been preached in every tongue, SIL's linguists aim to translate it into every language, speeding the destruction of those very languages by hastening the "last days" of the cultures they bully so viciously.

How else was the millennium celebrated? Befitting the liminal nature of any New Year, especially That One, come with me to a shoreline-city. Gisborne, on New Zealand's east coast, was the first city in the world to see the sun of the new millennium. (Appropriately for modernity's speed obsessions and overtaking desires, no one asked where was the last place to see in the millennium.) Here at Gisborne, though, now, at the edge of the sea, a little girl with legs like elastic is wave leaping and the cold water on her toes makes her wring her hands in the air—she is swept up in the shoreline moment of childhood, living both in now and in eternity. The sea, cerulean blue streaked with gold silk, is full of time and is the source of time—washing out an old day and washing a new one in.

The millennium was big on the Tourist Board's agenda in Gisborne. Very, very big. Even their post-office box was P.O. Box 2000. The organizer of the millennium events (the first of which was *Servant 2000*, a Christian gathering) was a middle-aged woman, navy-bosomed and as guarded as she was glossy. What does the millennium mean to Gisborne? "A time to go ahead. A time both to reflect back and look forward positively. It's a pivot point between past and future; it gives us a focus, a point on which to reflect." Her guard slips a little and she says briskly: "On another level, it's a marvelous marketing opportunity."

Would Maori people see the millennium as something to celebrate? An imposed Christian calendar, perhaps? "Oh, the Maori, I love them." A colonization of their time, a colonization of their land? "All land is important to Maori, so they own . . ." (she pauses, sighs, won't look me in the eyes but pats her bosom, speaking like a respiratory puff of talcum powder), "*spiritually* they own all places, lots of different things." Their material poverty, their poverty of land and their cultural impoverishment hangs like a bitter shadow behind her words. The bay here is called Poverty Bay.

Here is where it all happened in the year 2000. And here at Gisborne is where it all happened in 1769; here is where Captain

Cook first arrived to map and to claim New Zealand for the West. There is something burningly appropriate about the millennium being so celebrated here, the Western calendar of time-colonization being marked at a crucial site of Western land-colonization, an imperialism of time chiming with an imperialism of place—the enclosures of both land and time.

It's just before dawn, now; to see the sunrise you have to walk up the hill behind the city where a statue of Captain Cook stands, holding a sword and a set square, as if his landing, on October 8, 1769, were but yesterday. Cook's smug statue is gratifyingly graffitied; the base must have been repainted a hundred times. Someone has self-respectfully sprayed a swastika on the plaque with fluorescent orange spray paint and the orange writing, glowing in the half-light, says: "*Yo Fuck Yoza. M.M.M.*" (The Mighty Mongrel Mob, the largest Maori biker gang, claim Gisborne as their territory.)

Further up the hill is a whitewashed building, the James Cook Observatory, the World's Easternmost Observatory, marking degrees of longitude and latitude. It recalls the Greenwich Observatory and its special links with time and imperialism via the oceans of the world and the slave trade, and recalls Newton and the cruel light of his time-grinding mind. In Gisborne, some street signs pound this out: Gisborne, First City of the Sun, Longitude 178 degrees E, Latitude 38 degrees S.

Dawn lightens the whites' white observatory of white imperialism and Western time. Then something funny happens. You realize, for all the fuss about the easternmostness of Gisborne, it is actually very hard to see the sunrise, because the whole disposition of the city curls (fantastically aptly, considering its history) towards the *West,* not the East; it has, as it were, its back to the dawn and even at the highest and most easterly point, your view is obscured by factory-pines and fences everywhere. Fenced out, again, by the fences of settlers, the fences of enclosures, actual and symbolic, all the fences between the people, their lands and their time.

How else was the millennium celebrated? Sometimes wisely, sometimes well—a thousand-line chain poem circles the world; a hundred poets writing ten lines each, the poem picking up lines in any language. In Germany, Hanover hosts the Expo 2000—a global exhibition with the theme "Humankind—Nature—Technology," with the specific hope that, in future, technology should be put to better use than it has been in the past. In Britain, miles of new cycleways were planned.

Appropriately enough for the Christian millennium it is, Rome and Israel expected an influx of tourists. In Israel there were commemorative pilgrimages in Jerusalem and Bethlehem, and Megiddo was to have been put back on the map. Megiddo? You'd know it as Armageddon, which is what Megiddo used to be called. There was enormous interest from people planning to visit to "see the apocalypse for themselves" and entrepreneurs were attempting to build a biblical theme park there with a virtual-reality simulation of apocalypse. (It is at the moment the site of a high security prison.) The promoters expected some four million visitors. (Presumably not counting the armies of Satan.) However, the event was cancelled by the Israeli government worried about the potential hazards of apocalyptic fervor among some visitors.

In the U.S., a twenty-four-hour party was planned in Times Square and tickets for a party onboard a ship in California were offered for sale at a cost of two thousand dollars. Texan business-men bearing checks tried to buy the film rights to the last fifteen minutes of the millennium as it happened at Greenwich. (They were politely turned down.) In Lapland, an ice hotel was being built in the autumn of 1999 to house a millennium party—drink Absolut vodka and sleep on solid ice beds—before the whole ho-tel melted in the spring.

The Swiss didn't want to play. Sticklers for accuracy, they insisted that the new millennium wouldn't start until they said so; and their date was 1/1/2001. (Stephen Jay Gould pointed out that the millennium had in fact already passed thanks to our adoption of the Gregorian calendar.)

In France, there were lovely rumors of musicians floating in boats down the Seine in a liquid flow of melody so fitting, for the flow of waters, of music and the flow of time itself harmonize so. There were more definite proposals to pour *eau de toilette* together with two thousand plastic fish into the Seine. (Needless to say, the entire world is unofficially celebrating its rivers by continuing to tip poisonous chemicals into them. Business as usual.) Some of the French suggestions were a bit *demi-demi*; planting trees, which is nice, along the Paris Meridian in a dead straight line, which is not. Paris, which was witness to such key global-time events in 1913, was to see a massive lighting project turning *Place Charles de Gaulle* into a huge clock face and the Eiffel Tower "giving birth" to an egg, which seems an unusually female thing for such a phallic construction to do. There was to be a Ferris wheel like a giant transparent alarm clock, called *Le Chronos*. In the squabble of sibling rivalry that Britain and France have kept up for centuries, London got a Ferris wheel too. And, yes, the men in charge had a bit of a ding-dong over whose dong was bigger.

In Gisborne, a Millennium Clock, at the moment of the millennium, counted back from a thousand days to what they called a "full" display of zeros; though an "empty" display of noughts might be more correct, for how better to empty time of its meaning? Not that Gisborne is alone. Dublin authorities had the same idea of counting down to nothing-at-all when they disastrously put their digital clock in the river Liffey. On April 4, 1997, at the noughth meridian in Greenwich, at the so-called "center of time," the Old Royal Observatory, a clock was set up to count down the last thousand days. The makers claimed that it was the most significant timepiece in the world, because it is accurate to a millionth of a second—but this significance signifies nothing, it is a sign which empties time of its significance even as it, Kiplingly, fills each minute with sixty seconds' (and 60,000,000 millionths') worth of digital distance run.

And in all this emptiness of time, the most hollow thing was surely the Dome at Greenwich, "home of time." From the beginning, the Dome had one fixture, though not one the organizers advertised. A massive underground road, the Blackwall Tunnel, ran beneath it, and a giant exhaust pipe from the tunnel came up, a noxious fume-laden chimney, right through the dome itself. Certain things were from the outset planned to figure heavily in the dome's celebration of "time." Business. Commerce. Trade and Industry. Financial Interests. (All still peddling the same tedious lie, time is money.) Supermarket chains and car manufacturers were particularly avidly sought out for funding, which was really very curious, for those are the two industries which perhaps more than any others have polluted modern time beyond recognition. Arms manufacturers oversaw the "Mind Zone." McDonald's did the section on local culture. (Could this be the same McDonald's who, with their rampant global homogenization, are responsible for wholesale demolition of local cultures?)

The Dome was a symbol of modern culture, a horribly, hilariously appropriate self-portrait, for this age which thinks that time is money, which poisons itself on polluted air and holds arms dealers in high esteem, which is both empty and stuffed with overconsumption, which is enthralled to obsolescence—costing hundreds of millions of pounds, it was designed to last a mere 25 years. It is the aptest possible representation of society's treatment of time and of festival-time in particular. "Pageant" in its secondary meaning is an "empty show," and the pageantry of the dome is a self-portrait of a culture hollow but stuffed with overconsumption, emptying festival-time of its gorgeous variety, letting pageant grow paramount over festival, letting commerce dictate what community has forgotten, letting monolithic globalizing interests overtake the specially local, letting linear time hammer out cyclic time, allowing that bland indifference of supermarket-dometime to oust days of cakes and ale, the time of festival that is all but lost, the belching, shoogling, streamer-fluttering and transformative carnival-time, time of whirligigs, transvestites, of tickling, truancy and toss-pots, of pom-poms and lechery and drunken, tootling, vulgar horniness, festival-time mercurial, raucous, loud, rude as hell with bells on, with knobs on, and big ones. Big time. Gusto, give me excess of it.

5 · WREAKING GOOD HAVOC—
A TIME OF WOMEN

We are volcanoes. When we women offer our experience as our truth, as human truth, all the maps change.

—URSULA LE GUIN

"*Her* time of the month." Language-the-wise would call it her own; her womb knows her moon-time, her ebb and flow. Her own swollen hours are no one else's. (Though a masculine public calendar would rather blank it out.)

The moon swells to full, then a moon-shedding begins, ebbing itself to a crescent; the moon, symbol of women, fluctuates, alters, changes, flows in cycles. (Though a patriarchy would privilege the changelessness—of the sun—over the inconstancy of the moon and you.)

"*Her* time approached," they used to say, in childbirth. Hers. Yours? Mine? (Though a clockworked stopwatched public-time would cut it to fit a schedule.)

Woman, in her deepenings of age, lets the years write the lines of her story on her face. (Though a misogynist gaze dislikes those lines, preferring the empty page of youth.)

There is gender politics here, no mistake.

Beneath the bright white lights of the operating theater, male fingers move under the skin of a woman's face, tearing through the muscle and fatty tissue until you hear the thick knots of flesh ripping like raw meat. The nose and mouth act as staples to keep the skin attached to the face, just, but where the cheeks still cling, the surgeon's hands work, punching the tougher bits where the flesh still snags, until it is all flapping loosely enough for the knives to begin.

That's not quite what you see in the brochures for cosmetic surgery: they speak delicately of "incisions," where "the excess is removed" and the "skin is sutured into position." "To go under the knife is a pretty horrendous operation," says Peter Smith, spokesperson for one cosmetic surgery. "But the public don't want the gory details." Why do it? To be young. "This is a rejuvenating process," purrs Smith. "We're more youth-oriented than cultures in the past. . . . Women want to look younger, getting into the twenty-first century," he says, as if the transition to the next millennium were like struggling into shoes two sizes too small.

At another surgery, Karen, a client-adviser of indeterminate age, quite possibly fifty, recipient of a dolly-mixture of silicone and stitching attention, will let you look at the way plastic surgery has kept her "young," her looks are modeled so her breast-bubbles rise almost to her shoulders, she has that tight, fixed squeak of a face-lifted smile. She doesn't look young—she looks plastic. She doesn't look her age, true, but she doesn't look any other age either. The cosmetic surgeon consulted by Lana Turner remembers, rather sadly, how she showed him a photo of herself when young, and said: "I want to be like that." ("The secret to staying young is to live honestly, eat slowly, and lie about your age."—Lucille Ball.)

If man has seven ages, woman in contemporary western society has only one. One young one. One fixed one. Time must be stayed, for women, like plastic—with plastic. The terms "old man," "old chap" and "old boy" are tweedy, arm-round-shoulder terms, all clubabbly wheezing with shedfuls of common sense. Their female counterparts, "old woman," "old maid" and "old lady" all have horribly pejorative senses. Whereas men can be considered attractive well into middle age—the distinguished older man—women can only be attractive if they are—or seem to be—young. Women—old bags, old hags, old witches—with their saggy aged necks and baggy age-fallen breasts, should be hung by the neck, as witches were, to tighten the sags, tightening the noose of beauty around their throats. Cosmetic surgery has always targeted older women, but the age of plastic surgery patients gets ever younger—in the U.S., anti-aging surgery is advocated for women aged thirty onwards.

There's a price to be paid. A face-lift will cost around £5500, says one London surgery, others say more (and in Albuquerque, one surgery offers a face-lift at $4294.93) but there are more prices than financial. In a newspaper interview Helen Bransford, author of *Welcome to Your Facelift*, recalls how post-face-lift nausea induced vomiting that threatened to rip out her stitches. She had to throw up with her face muscles clamped like a vise. Surgeons warn of "infections, and scarring." The stitches, swelling and bruising, all hideously painful, last two weeks; a full recovery only comes after three to six months; and the effects will only last between five and ten years. Bransford was delighted with her face-lift, but her "before and after" pictures also describe a terrible loss. "Before" is older, certainly, but it is also the face of a powerful, deeply intelligent woman. *Her* face. "After" is an empty face, tame as glucose. *Anyone's* face.

The female body has long been considered faulty or defective. Aristotelian thought considered the male body perfect and the female imperfect while Leonardo da Vinci's "Proportions of the Human Body" used the male body (and quite an old body at that) to illustrate its supposed mathematical perfection. What is perfect is frozen in time, immobile and statuesque, and women's imperfect bodies today must be artificially perfected by surgery into a frozen statue resembling youth. It is telling that more than one cosmetic surgery uses statues on their promotional literature.

"It is important," says the literature, "to restrict motion of the face after surgery." The procedures of plastic surgery freeze the flow of time across the female face, fixing the expression, just as the purpose of surgery is to freeze-fix youth. The tightened face restricts signs of emotion like the stillness and fixity of a model's beauty. The picture (a *still*) on magazine covers shows both a stilled, fixed age, sixteen to twenty-four, and a stilled expression which expresses little. The very word *model*, too, tells of this life-less immobility. The face's whole meaning is a page to write your own character on, the whole purpose of having a face is to show emotion in motion—the mobility at the heart of expressiveness: but when age is not allowed to sculpt a face, but a surgeon sculpts it with a knife; when time is not allowed to play across the face with the sun-and-shadows of expression, but the client must hold her numb face stiff; what you get is neither beauty nor youth but the fixity of fitted plastic. Linoleum with lipstick.

Cosmetics themselves can have a similar effect. Lipstick steals your freedom of movement, so, wearing lipstick, you can't lick your lips with any gravy-gobbling gusto, you can't rub your mouth on the back of your sleeve, you can't pull faces with any self-forgetting success; but the delight of non-surgical cosmetics is you can use them whenever—or not—you like. All cultures have always used cosmetics to adjust nature to an ideal of beauty—the Georgians thickened their eyebrows with mouse fur, smeared their hair with lard and used toxic lead and mercury skin whiteners. Some societies use hoops to elongate necks like camels, others prefer tiny (rotten) bound feet, or lips stretched with plates, but our ideal has one repetitive criterion; youth, youth, youth.

We have moisturizers with "anti-time" devices "tested by NASA," and the Time Complex Capsules to turn old skin into young skin. . . . and we have an entire industry that, in order to justify its own spurious existence, must believe that the world is filled with women desperate to cling to their fading youth.   **137**

And we have just the one Anita Roddick, founder of The Body Shop, fly in the overpriced ointments of the entire cosmetics industry and speaker of that roomy, rolling rant. She, almost alone in the industry, is honest about age. She insists it is immoral to use a sixteen-year-old face to sell anti-wrinkle creams to forty-year-olds: she sells products to older women *as* older women. The Body Shop, she says, refuses to sell "youth." Because, simply, "it can't be done."

Women's age can not only be fixed from the outside by cosmetic surgery and cosmetic applications, but also from the inside. If you ask for Hormone Replacement Therapy, you may be offered a steroidal preparation called *Progynova* (meaning literally "a new woman," or "woman renewed"). *Progynova*, like an unction, is an oily word of slippery false hope; stay new, stay young. Other preparations are similar—*cycloprogynova, eugynon*. Though HRT can benefit certain medical conditions, its use can be interpreted as not so much treatment as symptom itself; symptomatic of a denial of all the ages of woman, the full female year, which must include the fall as it includes the spring. But for modern society, the fall of woman is a malady, a falling sickness, fallen breasts, falling wombs and falling-off cycles. HRT aims to make a supermarket of a woman, for like supermarket-time (always May, always mid-morning), so must a woman be always in the spring months, always in the morning hours.

HRT, the cosmetic industry and cosmetic surgery all work on women's bodies to stay the course of time. By them, though, women are denied—and deny themselves—the positions of elders; denied wisdom, power and respect so their "new" age has neither the girl-child gossamer of youth nor the grand seigniory of seniority; what they get for their money and pain is not youth but a pretense at youth, a patchy parchment promise which persuades just them. Like a no man's land, this is a no woman's time, a sort of temporal ha-ha where the joke is on us.

The eyes have it.

Gray is the least eye-catching of colors. Gray is the no-color, the no-shade neutral. Gray is the last-remembered, first forgot, the color last-hired, first-fired. Gray is the unhued sky which does duty only as background to rainbows. Gray is the dull emotion of boredom, gray is the color of depression, gray is the tone of absence and invisibility. No food or drinks are gray. While most colors are symbolic, gray is the one color to which no symbols cling. Don't go gray.

Especially if you're a woman. For, no longer noticed as blonde or brunette, your menopausal identity will be a gray-area, your gray hair will ensure you are ignored, missed, looked-over, unseen. "Older women are still invisible," says photographer Melanie Manchot, "and there is a real taboo around aging, particularly about showing older women's bodies." In her own recent work, she photographs her sixty-six-year-old mother, nude, in huge black and white prints. While the power of the (male) observer is, traditionally, vastly greater than the power of the (female) observed (the male artist and the female model for instance) and while this power difference is exaggerated by age—the porn "girl" and the dirty "old" man—something very interesting happens when, as in Manchot's work, the one looked at is old—the viewer is humbled and it is the observed, not the observer who retains the power. These pictures elicit an ancient respect for an elder.

When Karen claws her way into an age twenty or thirty years younger than she is, through painful and expensive surgical procedures, it is pitiful. When female TV presenters are valuable only within a certain age-range, it is irritating. When, by a profound cultural-physiological process, girls menstruate at an ever younger age (the youngest is seven), it is worrying. But when society dictates that small girls put on makeup and high heels to reach up into that same age of allure, sixteen to twenty-four, it can be tragic.

Kelly is gorgeous. Loads of curls, rigid with hair lacquer. Lipglossed lips in a fixed smile, eyes fixed with mascara and come-hitherness. And her satin dress is very short, her top very cropped. In one picture, she reclines, simpering, in frilly underwear. Kelly won the Mini Miss competition to find quite the prettiest little girl (and at the same time, quite the silliest little mother) in Britain. Kelly was eight.

Child beauty pageants began in America over thirty years ago. Today, incidentally, children are said to be subjected to plastic surgery to further their chances. The upside of the business may be nothing more than some queasy posing in the footlights and the tawdry tinsel trivia of parental vanity. The downside of forcing girl-children into an age-range far above them is this; pedophiles treasure the child beauty-contest magazines, to lust after the cute-and-knowing now-we-are-sixers. Six forced to sixteen, was the age of JonBenet Ramsay, American child beauty queen, when she was murdered late in 1996. The autopsy found possible evidence of repeated sexual assault; forced to look like an adult, she was forced into sex acts beyond her age. "Stay still for the cameras," she must have been told a hundred times and at her death she did not fight, she stayed still with the stillness—the awful passivity—of the long-term abused, till she became the epitome of stillness, the killingly assaulted dead child.

From "her time of the month" to "when her time comes" in childbirth to "the change of life," a woman's body goes through greater time-changes than a man's. The woman's *change* is usually more dramatic—either problematic or liberating—than the male menopause. Childbirth is inherently more momentous to the mother than to the father and the menarche is more sudden and symbolic than any male equivalent.

A friend and I once sat down together for half an hour and wrote out a list of premenstrual signs. Between us we got to fifty-two, some physical, many emotional, the acute sensitivity, the twitchy irritability. The list of "negative" effects is well known to women and to listening men; the *lacrimae rerum*, the clumsiness and the *ugliggr* destructive tendencies, they need no rehearsal. But what no one ever tells you is that these tricksy tides are so powerful. The much maligned *paramenstruum* (defined as the two days before a period and the first two days of it), floods you with insight, with surges of instinctual thought, with demanding intensity, with burning innerness, thinking at full feeling. It is a time when the world full tilts towards you; when you are charged with the electricity of thunderstorms. (During one of my first periods, I touched a friend's arm, and the spark between us burnt her wrist.)

Many cultures ascribe to the *paramenstruum* a sacred power. The word meaning taboo or sacred in Polynesian and Siouan is the same as the one for menstruating, and in Dakotan, the word *Wakan* means spiritual, wonderful *and* menstrual. In other societies, female shamans and witch doctors use the menstrual state as a source of their power. There is a Dogon myth, from West Africa, that a woman, finding the skirt of the Earth Mother, stained with menstrual blood, took it and wore it and it brought her enormous power over men. Eventually, men established domination over women, through stealing the skirt.

In Spain's Andalusia, women were "seen to have menstrual magic that can wilt flowers, kill bushes and trees, wound the backs of horses, and extinguish the fire in a lime kiln," writes Julian Pitt-Rivers. Also in Spain, a version of the Cinderella story is told in which Cinderella is given a "pelican suit" and is ugly while wearing it; she is called a "filthy little pelican pig-keeper" who "shits at the door." She is described as alternating between beauty and ugliness, switching between beautiful dresses and the pelican suit. When she wears the pelican suit, she brings about the death of domestic animals—a reference to women's destructive powers in menstrual magic, comments James M. Taggart in *Enchanted Maidens.*   **141**

(I would add that the feeling of shuttling regularly between beauty and ugliness is what women can feel shuttling between ovulation and menstruation.)

"Warfare is disguised menstruation," ran a slogan on the fence of the peace camp at the U.S. Air Force base at Greenham Common, in Britain, suggesting either that cultures that hate menstruation are the most bloodthirsty, or that men, not having such power humming in their blood, must mimic it externally in war. That the *paramenstruum* can be a time of great sexual desire, widespread research and anecdotal evidence shows. (Ask your friends, ask yourself: flings and love affairs begin so often at that time.) The Kachin people of North Burma have a word *majan* literally meaning "woman affair": it is used to mean love-song, warfare and the weft threads of a loom, combining the ideas of both sexual and warlike power. What of the loom? It is a widespread image of time in women's cycles, oscillating, back and forth, menstruation to ovulation. It is partly an archetypal image of how women traditionally spent their time, the repetitive weave. But Bruno Bettelheim interprets Sleeping Beauty's small bloodshed, pricking her finger on the spindle, as a girl's maternal inheritance of blood. The distaff side of the family is, of course, the mother's side. Homer's Penelope sits at her loom, shuttle going back and forth, love and war, war and love, menstruation and ovulation, ovulation and menstruation. (With symbolic appropriateness, one of the authors of *The Wise Wound*, the groundbreaking book on menstruation, was called Penelope Shuttle.)

Worldwide, the menstrual time is "the moon," or "the moment of the moon," Maoris call it "moon sickness;" the word menstrual comes from the Latin *mensis*, which means moon month. The moon's cycle, 29.5 days long, is also the average menstrual cycle. The moon, at full or new, draws cycles, affecting the sea's tides and woman's alike; the ebb and flow of the ocean, the egg and flow of the woman. Cross-culturally, the moon represents both magic inspiration and madness—like the *paramen-struum*. Witches, worldwide, have used the times of the moon to entrain their cycles, lying in the moonlight as they do in the Shetlands. In India, according to Mircea Eliade, witchcraft can be practiced only by moonlight and the moon is seen as the cause both of menstruation and of time itself.

Pirá-paraná mythology says that the moon copulates with menstruating women. To the Pirá-paraná people, the story of menstruation is the story of *Romu Kumu* the female creator. She was an immortal shaman, because she had a sacred gourd within her—the womb-shaped gourd of shamanic power, which, the story says, "stank of her vagina." When she lost her shamanic power to men, her menstruation became a punishment—she then had "fire" in her vagina and became an ogrish, sexually voracious woman with "fish-poison for pubic hair." (Don't mess with the premenstrual.) More seriously, this, like the Dogon myth, seems to echo the implicit history of women's experience; if the *paramenstruum* is respected, it is a time of exceptional power, but if suppressed it will become a punishment, a curse.

The menstrual cycle gives women a differing experience of time. While the contented *embonpoint* of ovulation is a good time for renewing subscriptions and writing sensible letters; the spiraling time of the *paramenstruum* is a good time for flaming arguments and making a bonfire of what bores you, for having good sex and for wreaking good havoc. At ovulation, everything is ticketyboo, tame and tepid, you can sort out mortgage details, fill in forms and be polite, but at menstruation you play with fire and know your own wildest, feral emotions, for the hour is incandescent, thought is quick, sudden as flame, this is the time of woman in her wildest and most isolate aspect. It rises, this feeling, like burning blood which cracks rocks and singes your lips, the rising tide of flame along the body's shoreline, so hot that only the moon itself is cool enough to soothe it.

Time itself changes. The sense of moment becomes acute; how best to use the molten energy of those hours seems critical. (It is when women are premenstrual that they find it hardest to choose what to wear, for in clothing, then, as in everything, time has a quality of criticality.) Each hour fills like a bowl flaming at the brim with dizzy momentness. The past casts a shadow over the present; wrenching nostalgia or terrible memories haunt you. There is a veering difference now between the woman's interior, idiosyncratic, private clock and the exterior public clock with its strict, regular beat. She is at a critical, cuspish catch of time, time colored, time with a character of change. The clock of masculine society is uncolored and uncharacterized, it is non-absorbent time, each hour the same, objective and linear, chronological as opposed to kairological. Just how different women's time can be is also shown in the way women can subliminally influence each other to bleed together in menstrual synchrony.

Masculine society places high value on people being the same over time, being reliable as employees, consistent as parents. Approval attaches to "being yourself" and disapproval to "not being yourself today." But the very word *self* implies a false singularity. Women are never one, they are at least two, at different times; when I'm ovulating I'm not the same as when I'm premenstrual. At one pole I may well be cooperative, relaxed and nice. At the other, I will be intense, difficult, powerful and unpredictable. (Probably.) Masculine society denies or penalizes this plurality of times. The menstrual cycle, write Shuttle and Redgrove, in *The Wise Wound*, is considered "a periodic illness only of significance as an inconvenient time-waster, irrelevant to properly lived masculine straight-line 'neat' life."

Masculine society seeks to homogenize women's time, by mocking, hating or ignoring that exquisitely timeful cycle, but many women feel an urge to take off, to run away from male time on these days. "Menstrual absenteeism" is considered a problem in the workplace of male time, but women find that much employment is too tedious to waste these weird and exceptional days on. Tribal societies often had a "menstrual hut" where women went not because they were banished by men, but because they sought a women-only place.

Trying to fit her inner calendar to the male calendar causes a woman's snarling premenstrual feeling to increase. But worse, it is a terrible waste of her strongest time, a waste of good *wodness*; far better to go with the powerful flow of its canted, slanted demands. Jung suggested women should have the first three days of their periods off work to escape the masculine time rhythms of work. Havelock Ellis said in 1910:

> The time may come when we must even change the divisions of the year for women, leaving to man his week and giving to her the same number of Sabbaths per year, but in groups of four successive days per month. When woman asserts her true physiological rights, she will begin here, and will glory in what, in an age of ignorance, man made her think to be her shame.

How many months are there in a year? It depends who you are. The male year is solar and counts months by the dozen. The female year is lunar and we have thirteen months, which is why, Bruno Bettelheim suggests, it is the thirteenth fairy of a fairy tale who offers the "curse." (It is also why this book, mainly written over a female year, has thirteen chapters.) Nils-Aslak Valkeapää writes in *The Sun My Father* that the Sami grandmothers taught how many months there were in a year and comments of invading Europeans that

> the guests had one month less
> they do not speak the language of nature.

Almost without exception, women vividly remember their first periods. For some, it is a rite of passage, a warm encandled moon mystery, such as modern witches will do a working for, a critical womanly time for a new woman. Oglala Sioux Chief Luther Standing Bear recalled Corn Dances which a father would hold for his daughter's first menstruation. But for most, an over-masculinized society steals the sweet, especial magic of the menar-che and replaces it with a stain; a muddy stain in your underwear and a maddening stain in your mind. It is a lonely, silent, anti-cer-emonial humiliation, a soiling shame for which girls have to be toilet-trained a second time. How to twist behind you to see if the bulge of a sanitary pad shows in the seat of your jeans, checking the blood hasn't leaked onto the back of your skirt. How to go to the toilet smuggling tampons in your sock, how to wrap it up, or how to flush it away, how to wipe the blood off the seat before others see it. How not to let women's inner, red and Other time of blood be seen in the male world of white, clean time.

In a woman's world, the menarche and subsequent periods can have an enchanted, inspired—and communal—aspect; women's trysts with each other and with the moon. In a man's world, the menarche means the onset of a lonely madness and a lifetime of trysting with the porcelain moon in the bathroom. Its import as inspiration unrecognized, menstrual time which patriarchy has tried so hard and yet so unsuccessfully to flush away will just keep coming back as that horrible returnee, that disgusting periodic madness, that obstinate red floater, the tampon in the toilet.

Even in the time of *day*, menstruation is still put in the cate-gory of the repellent: in the U.K., advertisements for "sanitary protection" (that awful phrase, half-disinfectant, half-disgust) can't be shown before the nine P.M. watershed—the hour before which Things Offensive must not be shown. And, furthermore, these advertisements are not allowed to mention "blood." Well, water-shed-bloodshed, it's before nine today and—pfah—I don't care. I can say blood. I'm bleeding now. I can feel the blood tugging down from my womb. There's blood on the page. *Mind your sleeve.*

Don't cross me now, for your own sake—and for mine. I am capricious, chancey, chaotic and unpredictable. I feel changed. I think in wild hatreds and wilder loves. Yes, I feel mad. Old memories I hate haunt me now. I've been crying. I feel both vulnerable and ragingly powerful. I look frightful and I feel like running away, flying. I want my female friends. I feel like only fire is a good enough image. There is stormy weather for an emotional weather report and red, red rain. I am crackling with that electricity which burned my friend's wrist, the electricity of lightning. I feel brimful, brinkful, womb like a vessel, like a cauldron, bubbling with hot dark liquid. I feel, in short, witchy as hell.

And near me, so they say, the very bees in the hives die. I can sour wine, wither corn, blast and burn away herbs and young buds. I can turn a looking glass dim, and cause "a sword, knife or any edged tool; be it never so bright to wax duskish . . . Iron and steel presently take rust and brass likewise, with a filthy, strong and poisoned stink . . ." So saith Pliny the Elder in his *Natural History*, on the power of a menstruating woman: "Hardly can there be found a thing more monstrous than is that flux and course of theirs." Well, no. It's more majestic than monstrous, more mysterious than disgusting and its burning, volcanic energy is more immense than Pliny ever knew. That Pliny died because of just such a burning volcano gives me a certain mischievous pleasure. (But only when I'm premenstrual.)

As Shuttle and Redgrove argue, the hatred of witchery is actually a hatred of menstruation, the power ascribed to witches is that of menstruating women, all the images are shared, the blood, the moon, the power, the heat, the vessel—womb or cauldron— the fire. It is because women can "come into their proper powers by understanding their menstrual cycles that it has been denigrated ceaselessly by men," specifically, men of the Christian church, in a society where male knowledge and science feared and loathed female knowledge. Traditional midwives were often identified as witches—a high proportion of the women arraigned in the witch-hunts of the sixteenth and seventeenth centuries were midwives. The witch-hunts coincided with the male take-over of midwifery when male science, hand in medical glove with the patriarchal Christian state, persecuted the "wise woman's" craft which involved a deep understanding of women's time, of her cycles, moons, conceptions and births.

Women's time, in all its senses, was overruled by men. Old women were hated and the legacy is with us still in cosmetic surgery's implants of silicone fear into female minds, while elder female faces must be plasticized into facile facsimile face-lifts. Women's inner, menstrual time is still treated with visceral disgust, making her lustrous, powerful time of the moon a time of shame and hatred. Women's time of childbirth was overtaken by male gynecology and the convenience-cesareans of today still testify to that usurpation.

"Each week I attempt the [*Observer's*] Everyman crossword—and fail miserably," Sarah Horan of Bolton in Britain wrote recently to the *Observer*. "In the last two years I have completed it successfully twice. On each of these occasions I have, within hours, given birth. Is there any medical research into heightened intellect as a sign of the onset of labour?"

Probably not. If men gave birth, for sure there'd be a million such studies, but let that pass. Her letter is a sharp detail in a larger picture; pregnant women have a powerful, inner time, especially at labor. That childbirths are linked to the moon's phases has long been known to midwives. That time is also deeply idiosyncratic is respected by the best of doctors. But women's time in childbirth has been overruled by men's coercive public time since the male takeover of midwifery, dictating that women should conform to the solar calendar, not the lunar and the public clock, not the inner, private one.

"Her time approached" is one of the most telling ways of describing the onset of labor; like the menarche, it is a critical time, but this natural and idiosyncratic time, *her* time, not anyone else's, is too often overridden by the schedulizing of childbirth for the convenience of doctors and hospitals, in the masculinized world of midwifery which will force time with forceps when a woman is slower than the public clock says she should be.

A hundred years ago, the safest treatment for slow labor was said to be "tincture of time." Waiting. Even the meaning of *expectant* mother means waiting, but today to wait is actively discouraged. For obstetricians, most of the measurements of labor are time-bound; contractions timed, premature labor halted, postmature labor induced and cervical dilation measured against time. Obstetricians speed up labor with vacuum extraction or cesarean section.

In her study of maternity care, *Forced Labor*, author Nancy Shaw shows how women giving birth are put, as it were, on an assembly line, governed by "industrial time." "A key feature of the industrialized management of birth is the imposition of industrial or clock-time upon the process of birth." According to midwives in Ireland, since the 1980 publication of *Active Management of Labour*, obstetricians have been obsessed with the length of labor. This book advocated that all women should have their waters broken on admission to hospital, that deliveries shouldn't last longer than twelve hours to speed through-put of patients and that slow labor be speeded up with drugs. In its wise etymology, what does *obstetrics* mean? To "be present," to "stand at." Not to speed labor but to be present at it. Not to force a woman but to stand by her.

Asking the U.K.'s Royal College of Midwives anything about women's time in childbirth was clearly a big mistake. They mocked the very mention of the moon. Asked for comments on the importance of respecting a woman's own time in childbirth, if, they said, they were given a written set of questions, the—male—heads of the PR department might, for a fee, ask a—male—midwife to give a quote. Their time is money, they said. Ah. The male-dominated and money-oriented medicalization of midwifery gives women's time short shrift indeed. In America, thank heavens, some people see things differently.

"Time is *the* most important thing in birth," says Jan Tritten, editor of America's *Midwifery Today*, practicing midwife for twelve years and natural childbirth advocate. What does the phrase "her time approached" mean to her? She mulls. "If you listen to that phrase, it has to be old-fashioned. It has a very sweet ring to it, her time approached; a woman wouldn't be allowed that now. The phrase you always hear is 'Oh-she-didn't-go-fast-enough. Oh-she-didn't-go-fast-enough-so-do-a-cesarean-section.'"

Doctors, she says, "put a woman on time limits. Instead it's better to let her go completely on her own time schedule. A woman will often go to sleep, for instance, in the second stage of labor, if she's allowed to." On the one hand, she says, "there is the time-protocol, the public clock and the *same* calendar, but a woman's own rhythm is very different. This is the first time we've *not* respected the idea of 'her time approaching.' The first time we've manipulated a woman's timing."

The U.K.'s Royal College of Midwives are unlikely to keep a chart of the moon's phases on the wall. Jan Tritten says the moon "*must* be important." She describes once having eight pregnant women who were due to give birth over a five-week period. Then there was a small eclipse of the moon and over a thirty-hour period, fifteen hours before and fifteen hours after the eclipse, all the babies were born.

No one knows when I was born. My mother, bless her, lost track of time. It was, she says, absolutely irrelevant whether it was

Tuesday or Sunday, noon or night, December or June—the objective public clock meant nothing to her when her inner time was so powerful. She gave birth at home, not coerced by any artificial clock-time, but for many women the two types of time are in profound conflict; the woman's unsurpassed sense of archetypal-time happens in a location more clock-dominated than almost any other, a hospital delivery ward.

Meg Fox, lay midwife, mother and feminist scholar, writes of time in childbirth: "The woman in labor, forced by the intensity of the contractions to turn all her attention to them, loses her ordinary intimate contact with clock time. For her, time stands still, moments flow together . . . in place of sequence, and linear relation, there is an overwhelming richness of sensation, which pulls her attention from the outer world. She is immersed in the immediacy of her experience . . . Absorbed in the rhythm of labor, in the work of her own body she is in touch with a truly timeless present . . . a realm beyond time, an experience of immortality."

Women in childbirth have a privileged access to a time unutterably different from standard-time, for procreation is in a sense "time-generation," women literally creating time itself through giving birth. But motherhood, this huge, archetypal work of creation, is denigrated in modern society. Instead of being recognized for its awesome beauty, the pregnant body is regarded as ugly, the act of birth is often a cruelly undignified medical procedure, and "being a mother" means taking the lowliest social status.

Why so? In part because of society's gendered attitude to time. Traditional women's work, including the work of motherhood, is cyclic, it must be done over and over again, characterized by repetitive chores, fetching water (which gets used), washing clothes (which get dirty again), cooking (for families who get hungry again). "Few tasks are more like the torture of Sisyphus than housework, with its endless repetition: the clean becomes soiled, the soiled is made clean, over and over, day after day," as

Simone de Beauvoir said. "I hate housework! You make the beds, you do the dishes—and six months later you have to start all over again," as Joan Rivers put it. Its cyclical, repetitive nature is illustrated in the word *char*lady, "char"—and indeed also "chore"—comes from an Old English word *cerr* meaning "to turn."

In *Cartas a Una Idiota Española*, there are expressive cartoons showing a woman endlessly slaving for a man, rushing to the point where, harried by hours, vexed by minutes, not only does she wear two watches, but also her face has become a clock, the hands turning and turning, faster and faster.

Woman's work is perishable work, including creating mere, non-durable, human beings. Men's work—linear, lasting over time, massive developments, one-off rather than repetitive—is considered much more valuable. (Motherwork is demeaned, the work of nuclear physicists is highly prized, though one creates time and life, the other toxicity and death.) The difference is over something much more subtle than whether a woman is paid the same as a man—though that is important—the difference is that masculine-style time is privileged and given high status while feminine-style time is unprivileged and accorded low status.

Language recognizes the cultural differences between men's time and women's time: the words "fathering" and "mothering" are not equivalent. "Fathering" is used to refer to a brief moment of conception, and paternity almost as possession: "mothering" is used to mean the long slow process of nurturing over years.

The gender difference in work patterns can be seen in hunter-gatherer life where men do the occasional, spectacular work of hunting, whereas women are more likely to do the tedious, repetitive, daily work of gathering. Beth Anne Shelton writes of the gender difference in work, noting that although American men have increased the number of hours they spent on household chores between 1975 and 1987 (Male Laundry Hours up tenfold from .11 to 1 hour a week, Male Dishwashing Hours likewise up from .27 to 2.3), yet the type of chores often varies, men doing more of the outdoor tasks which are, crucially, more discretionary, at least with respect to scheduling. You can put off mowing the lawn, for a day or week, but not making supper for a hungry child. She also comments that when there are children in a household, the data suggest that their demands on men are primarily financial while their demands on women are for time.

Men build skyscrapers. Once. Forever. For a whacking big fee. Women clean them, over and over again. For a pretty low wage. Male-dominated work is time phallic as a skyscraper, linear and lasting over time, and dry, dry as a drawing board; while female-dominated work is cyclical, repeated over time, the vulvic round wet work, the mop 'n' bucket. In Cameroon, gender is organized on the lines of the concepts wet and dry. Women, all flow, are wet; men are dry. In English playground slang, of course, children considered not physically or emotionally tough are called "wet," with overtones of effeminacy. Margaret Thatcher dismissed as "wets" those in her cabinet not as rigidly male in their outlook as she was. In the rituals of the Tlingits of Alaska and British Columbia, wetness is associated with impermanence, and sharply contrasted with dryness, which is identified with the eternal: as this chapter has shown, there is a widespread cultural perception that the eternal is "male" and the impermanent "female."

Take a situation in India. Women collect water, in some parts walking miles to do so; it is centuries-old work, flow-work, done and done again, to keep the whole river of life flowing. One of the most valuable jobs there is. Huge, masculine development projects are underway, though, building dams, massive, spectacular, one-off jobs. Each will dislocate and make homeless thousands of people and among other things force women to walk further and further for their water. The dam is "spectacular," while women's cyclic work is "invisible." Vandana Shiva, Indian feminist, physicist, philosopher and lay genius, writes:

The more effectively the cycles of life, as essential ecological processes, are maintained, the more invisible they become. Disruption is violent and visible; balance and harmony are experienced, not seen. The premium on visibility placed by patriarchal maldevelopment forces the destruction of invisible energies and the work of women and nature, and the creation of spectacular, centralised work and wealth.

Maybe it's all in the genes. To be male is to dam. To be female is to flow. Women flow in their courses and create time in the course of childbirth, creating the time which flows like a river across the world in the universal analogy. ("Everything flows," as the ironic, iconic Ionian Heraclitus said.) But patriarchy dams the flow, forcing mega-dam projects in the countries of the South, damming the rivers' natural flow. It is women who have led the resistance to the dams. On one celebrated (and full moon) night in India, women led ten thousand villagers to stop the building of a dam on the Narmada river. This is a conflict which encapsulates a far wider struggle—between patriarchy and women—a conflict of the dam-principle versus the flow-principle.

Time flows liquid as a river, but modernity's over-observance of the clock (the dam on the wall) stops the flux of time. Women's natural flow of time is dammed by cosmetic surgery into the unflowing look of women with the fixed-smile fixture of a fitted face. Male-dominated society damns old women, and the time of the menarche, damming the riverine time of childbirth into scheduled courses, both damning and damming menstruating women. Female time flows, in flux, in flood, in all its courses. Let it flow.

6 · WET ROUND TIME AND
DRY LINEAR TIME

"If you knew Time as well as I do," said the Hatter, "you wouldn't talk about wasting *it*. It's *him*."

—**LEWIS CARROLL**, Alice's Adventures in Wonderland

The Mexican Day of the Dead is a kind of *Danse Macabre* at full tilt. A funky skeleton dances a fantasy fandango, flowery sprigs on its bones and apples sprouting from its shoulder blades, insects playing in its skull, birds on its fingerbones, ivy on its thighbones and daisies on its toebones—the *reductio ad absurdum* of a hippy. Nothing so graphically illustrates the chirpy belief in time moving in cycles and in the reincarnations of life cycles; out of death comes life. Some of these skeletons are on bicycles, literally pedaling cycles, time freewheeling through life and death, *dem bones dem bones gonna ri-ide away*.

Is time an arrow or a bicycle, a straight line or a circle? Once, time was widely seen as cyclical; the Hopi image of time is a self-contained wheel, the Gabra peoples of East Africa have the idea of *finn* meaning fertility or plenty in the cycles of life and in Hindu thought, time moves in the unimaginably long cycles of the Kalpas. In the Aions of the ancient Greeks, eternity wheeled round over and over again, while the Stoics believed in the eternal regeneration of the cosmos. Aristotle said "for even time itself is thought to be a circle" and Plato described time as a "moving" or "revolving" image of eternity. Throughout history, time seems to have been thought circular since it could not be separated from the cycling motions of the sun, moon and stars.

The modern Western view of time is linear, moving like a   **155** ruler straight from past to present to future and in this it is highly unusual. G.J. Whitrow, expert on the philosophy of time, says: "Our conception of time is . . . exceptional . . . is one of the peculiar characteristics of the modern world."

This chapter does not offer a "brief history of time." It doesn't attempt to state what time *is* or *is not*, as religion or philosophy or science might do, but rather looks at how views of time—particularly in religion and science—are part of the whole cultural landscape. Specifically, it argues that, although seldom described as such, time has always been a highly genderized concept; linear time is phallic, male in shape, cyclical time is yonic and female in shape, as women's bodies have cycles. Further, the way time is pictured, or described, in any age, mirrors remarkably closely the way the feminine is treated then.

It is not easy to gauge whether women had respect, equality and power in the earliest societies which saw time as cyclic; ancient farming societies, for instance, to whom time was the agricultural cycle of seasons, the wheel of life. It is also difficult to say, for example in India today, that widespread images of time as cyclic parallel respect for women. It is too clumsy to regard the-position-of-women as an index to the subject, since gender in society is a subtle thing, and not necessarily coterminous with women and men. In India, for instance, the whole of society seems more feminine *in tone* than in Europe. What is far clearer is that with the arrival of linear, masculine time, the position of women is made lowly indeed and things feminine are denigrated.

Time, on either side of lifetimes, before your birth and after your death, has often been viewed as female; the mother's womb and burial in "mother" earth. The traditional gatekeepers at both doors were female and the midwife who ushered you in, washed and wrapped you at your birth would also wash and wrap you at your death.

In contrast to the modern West, which vastly privileges the

sun over the moon as primary time-giver, early societies also used the moon, with all its female associations. The moon, lunar lurer of time and tide, draws the floodtides of the sea and the bloodtides of the womb. The two characters of time, illustrated by the moon on the one hand and the sun on the other, are gendered—the moon changes over the course of a month, from full moon to new, corresponding to women's time experience over a month, while the sun stays the same shape, just as a man's experience of time does.

Worldwide, the moon is regarded as female and in most societies—though not all—the sun is considered masculine. As matriarchies gave way to patriarchies, so calendars based on moon-time became less important than those based on the sun. *Mon*day gave way to *Sun*day. Sun-worshipping warriors of invading Indo-European tribes vanquished moon-calendared, earth-centered civilizations, bringing their sky gods, warrior cults and a patriarchal social order. To the Huns, the word *Tengri* meant both god and sky. So too the word *Akuj* to the Turkana people of northwest Kenya means god and sky. They were both blue gods and mono-gods. Nomadism, points out George Monbiot, has a lot to do with this—nomads characteristically worship the sky and are monotheistic, the ancient Israelites, for instance, believing in one god living in a sky-blue firmament. Though their own lives were nomadic, their god was fixed and stable. Blue, widely symbolic of eternity, is the color which most represents this changelessness.

Mithras was the "Iron John" of ancient times, bull-slayer and warrior. As Sun-god, Mithras clambered to a sky-high prominence, beaming enormous rays of influence over the Romans—particularly soldiers—of the third century C.E., so the week's first day ceased to be Saturday but became *dies Solis*, day of the Sun, Sunday. Sunday was co-opted by early Christians as the day of the Lord, whose halo was the sun's rays. The Emperor Constantine, who so influentially converted to Christianity, in 321 C.E. called Sunday "the venerable day of the Sun." In the fourth century, the dating of Christmas on December 25th was dictated by the sun, for that was the date of the sun's new life after the winter solstice.

After the debasement of the—female—lunar calendar, the whole history of calendar-making became a very male affair and time measurement was harnessed to patriarchal power. It is men who have set the calendar, whether the Roman senate renaming July and August for Julius and Augustus Caesar, or Pope Gregory XIII introducing the Gregorian calendar in 1582. It's a male-thing, this, ancient and modern, from the Pope to the Pirelli calendar and the pin-up "flavor of the month."

Digression: true to the same matrix (or perhaps patrix) of ideas; patriarchy, plus solar calendars plus a warrior society, take a contemporary example. In Russia recently, manufacturers of the MiG jet fighter planes used a Russian calendar-girl (with the *nyet*-scowl that only Russian models would favor). "Our marketing gurus tell us that the sight of a beautiful girl touting jets will boost sales," says a male representative of the manufacturers, in language that would hardly take a Derrida to deconstruct. "So we made a calendar. Beautiful girls for every month of the year. We also have pocket-size ones." (I'm sure they do.)

The changeover from ancient, female ideas of time, lunar in calendar, cyclic in nature, to the modern idea of time, more solar in calendar, linear in shape and masculine in character, was heavily influenced by Judaeo-Christianity. Matriarchal religions cleaved to the idea of rebirth; early patriarchy, with its successive dynasties and its father-son genealogies (and Salma begat Boaz and Boaz begat Obed and Obed begat Jesse) rules out rebirth, re-placing its implied cyclical time with the linear time of linear descent. Patrilineal.

The Mexican Day of the Dead, while paying lip service to Christianity, is actually a survival of pre-Christian belief; time cycles on, one foot pedaling life and one death, in a round of rebirth which, like most cyclical images, has a redemptive swing. Christianity would stop it dead, push a stick in the spokes of the wheel. Like Zoroastrianism and Judaism, in its linear sense of time, early Christianity condemned cyclical time with vigor. St. Augustine in *The City of God*, wrote that "The pagan philosophers have introduced cycles of time in which the same things are in the order of nature being restored and repeated, and have asserted that these whirlings of past and future ages will go on unceasingly." By contrast, he argues, "It is only through the sound doctrine of a rectilinear course that we can escape from I know not what false cycles discovered by false and deceitful sages." The history of the universe is "single, irreversible, unrepeatable, rectilinear," unfolding as a "unidimensional movement in time," from the creation to the life and death of Jesus to the end of the world. The crucifixion, described as a unique, historical event, implied linear time.

Religions which see time as linear—phallic in shape—are those which are patriarchal—phallic in character. Zoroastrians had their male god Mazda, their divine kings and their magi, wise men. The Judaeo-Christian bible is a handbook for patriarchy and much of the New Testament was written by the misogynist Paul. In the *Talmud*, women, together with children and slaves, are exempt from the time-based "thou shalt" commandments because since they are not free men, their time is not theirs to give.

The figure of Jesus (if indeed he ever existed) as represented in the New Testament has a lovely tilt on him and is the sort of person a woman might want to sit next to at heaven's trestle table, but he was cut out of the patriarchal planning process. Christianity's triumvirate (God the Father, the Holy Ghost and St. Paul) is all-male. Its apostles, disciples and priests are male. For born-again Christians today, the masculine rebirth in Christ is more important than your original—female—birth. Even the shape of the cross, Christianity's most enduring symbol, uses masculine straight lines as opposed to the yonic circles of earth religions. In Somerset, England, a "circle calendar" of standing stones was recently set up by people living on the land, to mark the turning wheel of the year. In fury, nearby Christians set up a cross to "ward off the evil" of the cyclic honoring of "mother earth."

In Christianity there are profound parallels between the masculinization of time and the church's suppression of the feminine in the widest sense. The sixth century Rule of Saint Benedict was fundamental in altering perceptions of time. Historian of technology Lewis Mumford writes of it: "under the rule of the order, surprise and doubt and caprice and irregularity were put at bay. Opposed to the erratic fluctuations and pulsations of the worldly life was the iron discipline of the rule." But Mumford's point can be taken a (genderized) inch further; the masculine church militated against the very aspects of time which can broadly be described as *female*, moving in fluctuations: time capricious, elastic and changeful was to be eradicated by rigid, straight—masculine—time.

Both these ideas, the male domination of women and the masculine linearity of time, have become so successful, have so fully penetrated society's world view, that they can seem inevitable; having the cultural invisibility of taken-for-grantedness. But just as the male oppression of women didn't happen without a struggle, nor did the linear idea of time so easily overpower the cyclic.

In the Middle Ages there was a conflict between the cyclic and linear concepts, though the masculinizing of time would slowly win. How long an hour was in, say, the seventh century would depend on the season; an hour was a long, languorous lean in summer, but a short snap in winter. But once time—in Britain at least—had been divided into twenty-four strictly equal hours, the round, variable and stretchy yonic hour was but a nostalgia.

•   •   •

On June 11th, 1594, the poet Edmund Spenser married Elizabeth Boyle and his poem for the wedding, "Epithalamion," describes the whole day, from the birdsong of early dawn, the bridal dressing and the minstrels and the crowds in the streets, to the feast afterwards:

This day for ever to me holy is,
Poure out the wine without restraint or stay,
Poure not by cups, but by the belly full.

Spenser might have loved the wedding, but he counted the hours, longing for the light to go ("How slowly do the houres theyr numbers spend?"), for day-time to be over, to take his beloved into his sonnetteering arms and come together in the hushy dark of an Irish twilight.

The whole poem offers a breathtakingly sophisticated portrait of time. In the last four stanzas, "time" runs like a litany: "Send us the timely fruit," "Our tymely ioyes to sing," "And for short time an endlesse moniment." The sun's movement is described and the circle of the year drawn. The twenty-four stanzas represent hours, but, further, the lines can be counted either long or short, long representing the duration of time, short the division of time, while the sum of the long lines is 365, the days in the year. The bride is compared to the (female) time symbol, the moon, while the poet/groom identifies himself with the (male) time symbol, the sun.

It was a marriage on several levels indeed, for Spenser marries time cyclic and time linear. Nature's cyclic, seasonal time turns like a wheel, while the poem itself represents the linear, for written language is linear, containing the last lines of past time within itself, carving a narrative line out of the wash of experience. (I write this into the future: you read it into the past.)

Spenser's timing was immaculate for he wrote of the disappearing world of cyclic time at the advent of classical science's most rigid, masculine assertions of time's absolute linearity. It was a metaphoric "marriage" of the two sexes of time, the last marriage before the vicious divorce of Bacon's ideas, the Cartesian split and the Newtonian separation. The treatment of time and the treatment of the feminine are so culturally connected. Women's darkest hour at the hands of Christian misogyny: the hatred of women, women's knowledges, their power, midwifery and cyclical experience of time, came with the violent murderous hatred which lasted between 1484 and 1640 or so—the witch-hunts. Then came the final defeat of cyclical time at the hands of classical science.

Overwhelmingly, the prominent figures associated with chronology and the study of time are men, from John de Dons, master clockmaker of the Middle Ages, who took sixteen years to make one particular astronomical clock (which, after his death, no one could repair), to clockmakers extraordinaire Christiaan Huygens, John Harrison and son, astronomer-timekeepers Cassini; father, son and grandson. Few women make clocks, even fewer make philosophies about them—only recently have women approached the subject of time; such as sociologist Barbara Adam and feminist academics Frieda Forman and Caoran Sowton. Men, though, have always found the subject irresistible, from Aristotle's "time is the number of motion," to Einstein; from Augustine's question "What is time?" to Newton's absolute certainty of answer; from Kant to Kierkegaard, Spinoza to Leibniz, from Heraclitus to Hegel, Heidegger and Hawking.

But rarely has time been more important to the male mind than in the rise of classical science; an intellectual period which self-consciously genderized itself masculine. A period which saw women's knowledge as dark and dangerous—the knowledge of the moon—and men's knowledge as being of the "light," of the Enlightenment, the knowledge of the sun. A period which dug into the flesh of the human body and found not blood or nature but clockwork. A period which looked at the universe itself and saw a clock.

The sires of science asserted they sought a "masculine" philosophy with "virile" powers, linguistically explicit statements of male domination. Francis Bacon, 1561–1626, scientist, philosopher (and "somewhat worm-eaten personality," according to humanitarian Albert Schweitzer) spoke viciously of "putting nature on the rack and forcing her to reveal her secrets," and of science's capacity to "bind Nature to man's service and make her his slave," in his significantly titled book *The Masculine Birth of Time* (1602). Scientific inventions do not "merely exert a gentle guidance over nature's course; they have the power to conquer and subdue her, to shake her to her foundations." The most violent images in the work of Bacon and his contemporaries are from the witch trials, the rack, the torture and the forcing to reveal secrets. In a passing aside in one essay "Of Marriage and the Single Life" Bacon writes that "single men . . . are more cruel and hardhearted (good to make severe inquisitors)." Bacon's oft-repeated authority for his attitudes was the Bible.

Two remarkable features of the thinking of early scientists are their genderized character and their insistence on the image of clockwork for the linearly predictable nature of the universe. The conjunction of these two ideas is highly significant, for just as women and women's knowledge had been brutally conquered, so cyclical, variable time was conquered and subdued by linear and absolute time. (A residual belief in reincarnation is thought by some commentators to have persisted in Britain until the time of the *Malleus Maleficarum: The Hammer of Witches*, in 1486.)

The image of the universe as mechanical clockwork, the overriding image of the age, is thought to have been first used by

Nicholas Oresmus (d. 1382) describing god as a clock-maker—this represented the first depictions of nature as a dead machine. The German astronomer Johannes Kepler (1571–1630) said "The universe is not similar to a divine living being, but is similar to a clock." French philosopher and scientist René Descartes (1596–1650), compared a healthy man with a well-made clock. Robert Boyle, scientist of mechanical philosophy (1627–1691), wrote that the universe is "a great piece of clockwork," and Bacon, in *The Masculine Birth of Time*, set out the new concept of linear *intellectual* progress. Time's linearity was supported by Leibniz, Barrow and Locke.

Cometh the hour, cometh the man.

Nature and Nature's laws lay hid in Night
God said, Let Newton be! and All was Light.

—ALEXANDER POPE'S EPITAPH FOR NEWTON

Isaac Newton (1642–1727) said the universe was clockwork. Moreover, he saw time as absolute and uniform: "Absolute, true and mathematical time of itself and by its own nature, flows uniformly without regard to anything external." Newtonian physics, the science of the mechanical, changed nature's "mystery" into "machinery," as Vandana Shiva writes: "The rise of a patriarchal science of nature . . . in Europe during the fifteenth and seventeenth centuries . . . transformed nature from terra mater into a machine." Previously, as Pope has it, nature was hid in night. That night-time of the moon gave way, absolutely, with Newton, to the time of the sun. Shiva notes how the effects of the witch-hunts which "aimed at annihilating women in Europe as knowers and experts" were compounded by the scientific revolution. As a result "women in Europe were totally excluded from the practice of medicine and healing because 'wise women' ran the risk of being declared witches."

The very word "wise" has a quality of time. Only age can be wise. Youth may be clever, but only age is wise. Newton had his *Annus Mirabilis,* aged twenty-four. Brilliant, absolutely, but wise? As Hegel says, "The Owl of Minerva comes at dusk," wisdom comes at the end of the day. The whole scientific revolution was a battle over two types of knowledge, the overpowering of female *wisdom* by male *brilliance,* women's *mysteries* of knowledge, as herbalism and midwifery were called, by men's *mechanics* of knowledge.

The age of "enlightenment" pitted itself against the dark aspect of the moon, and sought, with the full might of scriptural authority, the light. Light was privileged over dark, the visible over invisible. Female knowledge, irksomely, darkly subjective, was denied: it was *im*plicit, *in*tuitive and *in*terior, while the new science was *ex*plicit, *ex*periential, *ex*terior. While the interior is invisible, the exterior is visible. Visibility is a male affair; the cock stands *Ecce Homo* in the light. The yoni laps itself in the dark.

In the Enlightenment, the darkest sky was made visible by telescopes and the very nature of light was itself examined, by the white light of Newton's—brilliant—mind, and whose—brilliant—optical experiments made light's very elements visible. In 1693, the speculum as gynecological instrument first entered medical dictionaries: in name and nature an instrument for looking. The bright light of gynecology as a male science began to shine into that darkest place of all, pitch red and womb dark; at once the heart of the individual woman's mystery and the heart of women's science—midwifery—women's mystery. And that same bright light still shines on the female face under the cosmetic surgeon's knife and eyes, making her an object for men's eyes as much as the Russian *nyet*-calendar-girl "flashed" at by male cameramen. Women are to *be* looked at, passively, for to look is the prerogative of men since the Enlightenment. Wise women arraigned in the witch-hunts were in part detested for the "penetrating" quality of their gaze, as Germaine Greer writes in *The Change*, which, like the penetrating mind, was a privilege reserved for men.

From the Enlightenment on, what is visible has been privileged and science, technology and engineering have promoted "spectacular" erections, be they the Eiffel Tower or Newtonian telescopes. In honor of the phallic, penetrating instruments of looking, in 1750, two constellations were named Telescopium and Microscopium—and in honor of clocks, another constellation was called Horologium. Time itself was—supposedly—made visible by telescopes and made manifest in clocks, from sundials onwards, with the *gnomon* penetrating the round circle of time on the dial and the invisible hours of night becoming visible in twenty-four hour clocks. The fascination with the visible meant that the clock, objective and visible *machine* of measurement, became privileged over the subjective invisible *mystery* of time itself—just as it is today. Time mechanical, linear and phallic won the battle and clock-time (visible and spectacular) was considered *the* time. True to that set of ideas, a "spectacular erection" was built in 1675 to mark men's time-measurement, in its most visible, most observable sense, a palace to patriarchy; the Greenwich *Observatory.*

The scientists of the Enlightenment saw themselves as objective spectators. But what you see depends on how you look, as any quantum theory professor or student of human nature will tell you. No science has ever been truly objective and when classical physicists viewed time, they saw it in the image of their—male—minds; as linear, phallic, rigid and absolute. But others, looking differently, have seen otherwise. Mayan astronomers dug deep dark pits or holes in the "mother earth" like an inverted telescope to look at the night sky. And as the Enlightenment telescopes were phallic, and saw phallic, linear time, could you not argue that that is why the Mayans saw cyclic time through their deep-dug holes, truly telescopes of the cunt?

Time to scientists during the seventeenth and eighteenth centuries was absolute and deterministic. In the wider world, the increasing precision of clockwork (coupled with the increasing number of clocks and watches) meant time was chiseled to fit snug to the clock—the last seconds of inaccuracy planed away to a tiny curly shaving. Time must be predictable, knowable and visible—and, most importantly, time was considered synonymous with clockwork—as, all too often, it sadly still is.

As a result of all these things, time was stripped of its female nature—most obviously in the antipathy to cycles of time, but also in the tricksier characteristics of time as a thing of chance, caprice and unpredictability—all things which, for good and for bad, have been associated with the female. Aristotle used the terms "male" and "female" to describe differing understandings of time in the cosmos, calling the heavens male because he considered them eternal and immutable, while calling the earth female because it was changeable. "*Varium et mutabile semper femina*," according to Virgil (Women were ever things of many changing moods). So time, then, at this period in history, was considered not only linear, but also to have those same male characteristics (which the scientists prized in themselves) to be rigid, unfluctuating and absolute, not changeable like the female moon, but changeless as the—male—sun.

That *time itself* slipped utterly out of their overreaching clutches—like the cool disc of the full moon slipping out of the grubby hands of the little boys in the New Guinea myth, gliding smilingly up across the overarching sky—is the story of twentieth-century physics; the redemption of science.

What happened?

Deep-subjectivity came along like a sphinx with a banana

skin and tripped up the scientist in his march into the light. Fallen, with a sore elbow, bruised knee and ego, the pouting scientist to the sphinx: "Who are you anyway?"

"Entirely up to you," replies the sphinx, as particle. "I see! You are a particle!" cries the scientist. "What you see depends on how you look," murmurs the sphinx as wave. The scientist rubs his eyes and, yes, as if by some mysterious magic, the sphinx is a wave again, a mock-a-minute to stun the three-hundred-year-old certainty of science.

The world of science changed. Relativity. Quantum mechanics. Chaos theory. Dissipative structures. Heisenberg's Uncertainty Principle and quantum theory; whether something is a wave or a particle depending on how you look at it. Male objectivity caved in.

Science, during this century, has bewitchingly metamorphosed, into something richer, stranger and more *tender* than the rigid mechanistic science of classical physics. The very style of scientific terminology—the watchwords—have changed. Newton's "Absolute Time" becomes Einstein's "Relativity." Bacon's utter self-assurance and "certainty" becomes Heisenberg's "Uncertainty." Classical "determinism" becomes twentieth-century "chaos." "Necessity" gives way to "Chance." Where Kepler insisted that the universe is "not similar to a divine living being, but is similar to a clock," Gaia theory has gone to the heart of what it means that our earth is alive—its warm and moist life-giving properties. Disciplines which were self-consciously masculine are now shyly embracing the feminine. The stance of Bacon's coercive, violent torturer-scientist becomes that of more tentative sensitivity. Theoretical physics shows a greater humility in the face of all that is not known, as what, anciently, was mystery and became machinery, now admits of mystery again; the awe and the knowing no longer in opposition. It has been a history, if you like, of the "feminizing" of science.

A big disclaimer is needed here; I don't regard *masculine* and *feminine* as necessarily qualities of men and women respectively. They operate as cultural shorthand for broad accretions of ideas. For their non-alignment even within science, take, for example, two scientists, one sensitive, complex, holistic, ecological and deeply influenced by the feminine—Fritjof Capra. The other rigid, patriarchal, cruel, hierarchical, war-mongering and masculine—Margaret Thatcher. "Masculine" and "feminine" are a quick notation for a whole matrix of tendencies, like yin and yang in Chinese philosophy. (The yin/yang sign was the coat-of-arms of the Danish theoretical physicist Niels Bohr, with his motto of complementary opposites.)

In every culture, there is both "male" and "female;" in modern society, though, the male side has been too privileged for too long. Chinese philosophy teaches that whenever one tendency overreaches itself, it contains the seeds of its opposite tendency within it; just when "male" science reaches its apotheosis in erections of rockets, bombs, biotechnology and dam-building, so physics is also producing "feminized" science; the work of Capra, James Lovelock's and Lynn Margulis's Gaia theory, Ilya Prigogine's work on time.

In no subject was this "feminization" of science more intriguing than in the subject of time. When Ilya Prigogine surveys twentieth-century physics—relativity, quantum mechanics, evolutionary cosmology, non-equilibrium structures and deterministic chaos—he writes: "a remarkable point is that all this emphasizes the role of time." Much of Prigogine's own breakthrough work was being done during the 1970s. That period was also the decade *non pareil* for the feminist texts of the women's movement: *The Female Eunuch* setting the pace in 1970. And right in the middle of that decade, in 1975, two classic texts were published. One, by Capra, was *The Tao of Physics*. The other, by linguistic scientist George Steiner, was *After Babel.*

These two powerful minds at work on subjects a world apart, nevertheless reveal something very similar. Capra writes of quantum theory: "At the sub-atomic level, matter does not exist with certainty at definite places, but rather shows 'tendencies to exist,' and atomic events do not occur with certainty at definite times and in definite ways, but rather show 'tendencies to occur.'"

Compare that with Steiner, writing of time in women's language: "At a rough guess, women's speech is richer than men's in those shadings of desire and futurity known in Greek and Sanskrit as optative; women seem to verbalize a wider range of qualified resolve and masked promise." (Compare that also to the Hopi view of time where time is divided into the "manifest," curving to the past and the "about-to-become-manifest," with its list into future.)

The picture of time which Capra's physics and Steiner's linguistics each reveals could not be more different from Newtonian rigid, absolute time and what is revealed is *female* in character. It is the curved slant of the glancingly evanescent; like an invisible net of silk streeling away, cast from the hand of *now* into the unreached water of future-potential and, watching its arc wait on the rise of a moment, you can almost hear time—subtle time— breathe in.

Patriarchal thinking, from its sun-worshipping warriors of *Dies Solis*, reaches a pinnacle in today's over-solar *Dies Irae*, the brightest, ferocious white light of atomic warfare outshining the sun. The worst of masculine science has betrayed the world in its deadly refusal to be socially responsible; in its bombs and biotechnology, its phallic rockets blasting the planets' eternal cycles. The best of science, though, is a passion for life. Gaia theory expresses the gorgeous vitality, the "exuberant disequilibrium of the Earth" as Lovelock says. The term "life" runs through Lovelock's work like a jazz riff: life, lif, alive, life, anima, life-alive.

The life sciences, as Lovelock demonstrates, are overwhelmingly important both as disciplines and as politics; he includes passionate environmentalist pleas in his books. There is a detectable tendency—albeit still in a minority of scientists—to use multidisciplinary thought that is not shy of values and politics. In *The Web of Life* Capra writes: "The new concepts in physics have brought about a profound change in our worldview; from the mechanistic worldview of Descartes and Newton to a holistic, ecological view." Sociologist Barbara Adam and anthropologist Alan Tormaid Campbell are ahead of their disciplines in taking a moral stance in their work. *Science for the Earth*, edited by Tom Wakeford and Martin Walters, is a compendium of broad scientific thinkers, looking at the world with sidesight (wiser than hindsight), those who connect, who think with "all" halves of the mind and who consider it a duty that scientists should "make the world a better place," as the subtitle has it.

Thanks to feminism, there is also a new–but–very–old understanding that knowledge can be rooted in the body—that the personal is not only political but intellectual—and the body is not the opposite of the mind, but is itself a way of knowing. Danah Zohar, mother, physicist and author of *Through the Time Barrier*, discussing her experience of motherhood and comparing it to time in quantum physics, says: "These experiences of time—in dreaming, reflecting, being a mother, are not easily given to structure. When I was pregnant, schedules didn't sit well with me at all. Trying to structure the flow of experience into Newtonian sequential time gave me a headache."

In the best science today, there is a surge of thinking, half exasperated with the carpeted chuckles of single-discipline academia and half exuberant with the elation of thinking widely. Scientist Friedensreich Hundertwasser writes this wonderful long-jump of thought, leaping from the scientific concept "straight line" to its social effect. "The straight line leads to the downfall of mankind . . . It has become an absolute tyranny . . . something cowardly drawn with a rule, without thought or feeling; it is the line which does not exist in nature. And that line is the rotten line of our doomed civilization. . . ." Opposed to that straight line is the circle principle, of a worldview in which ecology and feminism link up: ecofeminists show how the treatment of nature has mirrored that of women—noting in passing how patriarchal capitalism has sulkily hated the very concept of recycling with all its female overtones.

And what of Time? Classical physics saw time as a separate dimension, absolute and independent of the material world; Einstein recognized that it was relative and dependent on the observer. Space and time seemed entirely disconnected in classical physics; in relativistic physics, they were unified. As the best science has been "feminized" in the twentieth century, the tendency has never been truer than in the subject of time. The force of gravity, according to Einstein's relativity theory, has the effect of "curving" space and time: Stephen Hawking's work showed the strong "curvature" of space-time around black holes and the infinite stretch of time over the event horizon. Time, instead of being rigid, linear and strict, began to be seen as elastic, stretchy and curved. Thinking itself had to reflect it; not for nothing are these ideas called mind *bending*, for they curve the brain and crescent it. Time, once stiff with the phallic principle, began in the twentieth century to curl into the yonic principle.

And then there are non-linear equations. British mathematician Ian Stewart writes: "As the world was a clockwork for the eighteenth century, it was a linear world for the nineteenth and most of the twentieth" but a decisive change has come about now in seeing that nature is "relentlessly nonlinear." With chaos theory, the universe is shown to be a long way from clockwork, a chancey thing, dynamical chaos laughing like a hyena at Newton and classical physics' time-symmetric determinism and predictability. Chaos is the rule not the exception. Chaos Rules, O.Kaos.

F. David Peat, an expert in the science of Native American cultures, says: "Now scientists are discovering what Indigenous science has long taught; that the norm is not order but chaos." Eighteenth-century mechanics considered time as reversible and, though common sense suggests otherwise, no model could be found to illustrate its irreversibility. Further, it couldn't countenance anything containing both movement and motionlessness in time. But in the 1970s and 1980s, Ilya Prigogine did an Archimedes with whirlpools and eddies and had his eureka moment. Whirlpools, eddies and candle flames were, he argued examples of things which contained both a structure and a flow. He coined the term "dissipative structures" to describe the presence of both ideas, shape *and* flux—which had seemed inherently contradictory opposites. The result of Prigogine's work with Isabelle Stengers was that finally you could see how in living systems, irreversibility ruled; it was the mechanism that brought order out of chaos and you had an image of time which included both stillness and motion, time arrested and time passing. Life itself is an example of dissipative structures.

Capra comments:

> The key to understanding dissipative structures is to realize that they maintain themselves in a stable state far from equilibrium. This situation is so different from the phenomena described by classical science that we run into difficulties with conventional language. Dictionary definitions of the word "stable" include "fixed," "not fluctuating," and "unvarying," all of which are inaccurate to describe dissipative structures.

Classical physics wasn't fond of flux. Arguably, this was part of the stone-blindness of classical scientists to all that is "female" especially in the shadings of time. No science is free of cultural conditioning, and hard-line classical physics combined with some hard-core classic misogyny to detest the very thought of flux. The period which saw the rise to power of classical physics also saw a spectacular hatred of menstruation—the individual woman's *flux*. In nineteenth-century fashion, a woman's exterior physical movement was hobbled in "hobble skirts," and even her interior movement was dammed: stays, laced too tight, as they often were, could literally stay—impede—the flow of a woman's courses— her menstruation. Well into this century, the hatred of women's flow is still apparent in the silence and shame with which the subject is treated. In the U.K., advertisements for sanitary pads, designed to illustrate the absorbency of materials, are not allowed to represent blood by a red liquid. They have to use its opposite color—blue. Red is the color of life in *flow*. Blue is the color of stasis, of motionless *structure*—blue as the sky habitation of the patriarchal god of the ancient Israelites, blue as the god of the Huns. Arguably, this explains why the Catholic church dressed the static iconic Madonna in blue—not for her the menstruating power of womanhood in red, red flow.

Perhaps it took not only the "feminizing" of science, but the whole cultural resurgence of the female during the twentieth century, most evident in the women's movement (another flow), to enable scientific breakthroughs such as Prigogine's, to see time once more as *both* flow and structure. Capra writes that the radical character of Prigogine's work was:

> apparent from the fact that these fundamental ideas were rarely addressed in traditional science and were often given negative connotations. This is evident in the very language used to express them. *Non*equilibrium, *non*linearity, *in*stability, *in*determinacy, etc., are all negative formulations . . . Many of the key characteristics of dissipative structures—the sensitivity to small changes in the environment, the relevance of previous history at critical points of choice, the uncertainty and unpredictability of

the future—are revolutionary new concepts from the point of view of classical science, but are an integral part of human experience. Since dissipative structures are the basic structures of all living systems, including human beings, this should perhaps not come as a great surprise. Instead of being a machine, nature at large turns out to be more like human nature—unpredictable, sensitive to the surrounding world, influenced by small fluctuations.

But I would take this observation one step further, not entirely capriciously; the shades of time's character, as revealed in the model of dissipative structures, have all the qualities which patriarchy has attributed to *women*, and has therefore devalued for so long. It is *female* human nature which has long been expressed only negatively, so where Capra cites "nonequilibrium," the cultural dictionary might say women are unbalanced. For "nonlinearity," the cultural lexicon might offer the idea that women are illogical; for "instability" substitute women's famous unstable natures; and for "indeterminacy," would it be accurate to say that women are indecisive? I dunno. Can't decide.

Certain aspects of time—like "chance"—have lately been given new status, and the intriguing thing is that all these newly respected new-found shadings of time's character are those considered female. Chance has always been ascribed a female nature; Fortuna, Lady Luck. Fickleness is what women do best. Caprice is female, as Predictability is male. (And, oh exuberant disequilibrium, how I know you.)

Last night I dreamt of dissipative structures. When I woke, I was thinking about what they call the bifurcation point. (Horrible word that, all buck teeth and spit. Sorry.) The bifurcation point of a dissipative structure is "a threshold of stability at which the dissipative structure may either break down or break through to one of several new states of order," in the words of Capra.

(You can doodle it. If you take a line across the page, then stop at what you can now call a bifurcation point, then doodle in the myriad of possibilities, you get a shape like a dandelion, childhood's most charming clock.)

Capra again: "At the bifurcation point, the dissipative structure also shows an extraordinary sensitivity to small fluctuations in its environment. A tiny random fluctuation, often called 'noise' can induce the choice of path." At this point, far from equilibrium, chance dances with necessity, fluctuation waltzes with determinism. You are a dissipative structure. So am I.

To recap a little; dissipative structures reveal a renewed status for "female" aspects of time. At the bifurcation point, these "female" characteristics are at their maximum. And what of women? At the points of maximum femaleness, such as menstruation, the dissipative structure which is me or you is at its bifurcation point. "Sensitivity to small changes in the environment," "unpredictability," and "influenced by small fluctuations," are phrases used by scientists to describe this point for dissipative structures. Remind you of anything? Being "open and far from equilibrium?" This sense of being at a critical time—painfully sensitive, capricious, changeful, unstable, powerful, chancey, fluky, fluctuating, unpredictable and all—is exactly what it feels like to be acutely premenstrual; far from equilibrium indeed—chaos theory in motion. When the social equivalent of a butterfly flapping its wings just one time too damn many will cause the emotional equivalent of an erupting volcano. When one fraction of a temperature alteration will cause a flood of hot flushing way out of control. And, true to chaos theory, it is out of the utterly chaotic turbulence itself, the flow of "courses," of course, that a new order—postmenstrual self-collectedness—re-establishes itself.

All this time, women had the model of chaos theory and dissipative structures tucked right up their skirts. All those men, making all that effort. All that thought. All those late nights and all those non-linear equations, for Christ sake, why didn't some-body just ask *us*?

For we are linear, and we are cyclic. Our lives move like a line: child, girl, maiden, mother, crone. Yet for much of our lives, time moves within us like a circle. State-of-the-art science now provides a model of time which matches this very female experi-ence. Coveney and Highfield sum up their meticulous book *The Arrow of Time*, thus: "What has emerged from non-linear dynam-ics and non-equilibrium thermodynamics . . . offers a sophisti-cated reassessment of time. Non-linear equations show us that thermodynamics can account for both linear and cyclical time."

A marriage is announced of two times, the linear and the cyclic, and Spenser's sun scrolls round at last. Sometimes it takes a poet to know what it needs a scientist to prove; so that what once was the poetry of image can become the poetry of fact.

Spenser, in his "Epithalamion," wrote:

Ah, when will this long weary day have end,
And lend me leave to come unto my love?

It is a lover's question for his beloved and a question of one aspect of time for its complementary opposite. For time, forced for a day that lasted three hundred years, to walk in line in strict steps along a cold stone corridor of too masculine a thought, has all the while longed for its complementary opposite—cyclic time, shunned and vilified, waiting outside in the half-light of the moon. Mod-ern science has just now allowed their meeting at twilight, the trysting hour, the cockshut hour, in a dusk which holds both night and day, for the model of time science now gives us is one of time united, male with female, linear and cyclic, a fulfillment of the deep erotic principle at the very heart of time.

7 ○ THE POWER AND THE GLORY

How does the little busy bee improve each shining hour and gather honey all the day from every opening flower? Well, he does not. He spends most of the day in buzzing and aimless acrobatics, and gets about a fifth of the honey he would collect if he organised himself.

—HENEAGE OGILVIE

The last laugh, as any child knows, lasts longest—remember the belly-laugh of the titch-supremacist colonizing the corner of the playground? "And don't answer back," adults say to children, reserving for themselves the power of having the last word.

Ordinary human relationships are suffused with time as power—who doesn't know that pert, delicious smugness of being first out of bed? To keep someone waiting is, in the West, an act of rudeness: time is used as an index of power. In language, the connection is explicit, waiters are those who wait, whose time is at another's disposal. Attendants (from the French *attendre*, also to wait), are those whose time is on pause for someone else. Those who are—or think they are—important, will not wait in lines or queues, for they are for the powerless. Defoe's Robinson Crusoe, to take control of his stranded situation, makes calendars and, in a famous act of power, finds a servant to dance to the music of his time, whom Crusoe christens with his own calendar: Man *Friday*.

No one wants to be "the last to know" any news and, by contrast, the frequency of news is indexed to status so VIP's must often be updated. A pedestrian may stop for a hearse, marking their respect for the dead with a special pause. To interrupt your own time, thus, is to offer respect; to interrupt someone else is disrespectful, as children are taught. But there is a power differential in the pecking order of butting-in, which is both age- and sex-related. Children, of course, of low status, can be interrupted. Many studies find that men overwhelmingly interrupt women, though they rarely interrupt each other, illustrating how interruptions are not inherent to men's speech-patterns but rather signal social power. One such study, from 1975, is reported in *Language and Sex: Difference and Dominance*. A further, related, study in 1980 showed that parents were far more likely to interrupt their daughters than their sons.

On a far larger scale, time is aligned with power, so rulers, governments and priesthoods have always used it to create or confirm their authority. From Druids, Julius Caesar and Pope Gregory to the legions of Jesuit missionaries, from Chinese mandarins, British colonialists and Stalin to the authors of the French Revolution, the calendar is an ideological, political and religious weapon. Elias Canetti said that the regulation of time is the primary attribute of all government. Potentates, princes and priests, all hypnotized by hopes of hegemony, have always stood on the borders of space and looked at time—they have come, they have seen, they have conquered—for time is a kingdom, a power and a glory.

A Druid priest, on first miscalculating the Solstice, would have been laughed out of henge. Their power—assuming there actually was such a thing as a Druid priesthood—derived from their predictions of time, for it is but a skip between secretly knowing when something will happen and appearing to have caused it. Druid priests used time to authenticate their power, "making" the sun rise over a certain part of the stone circles and thus seeming to control time and season.

Priests, as far back as history considers, have made the control of time their preserve. Mayan priests of Central America erected famous monuments to the passage of time; G.J. Whitrow writes that "of all ancient peoples, the Mayan priests developed the most elaborate and accurate astronomical calendar, and thereby gained enormous influence over the masses." In Babylonia, the priest-king performed the same ceremonies on earth as their god performed in the sky: the timing of the priest's ritual had to chime with the god's. Worldwide, shamanistic rainmakers, shaking pebbles to whip up a naughty little shower, or hurling boulders for a torrential downpour, would have earned little kudos unless they could control not only the fact of rain, but also the time the rains came. In the Old Testament, when Joshua wanted to demonstrate power, he made the sun and moon stay still, while Isaiah moved the sun back ten "steps" on the dial.

At the 664 Synod of Whitby, the date for Easter was much discussed—a seemingly trivial but actually potent point. The question was whether Easter should be dated according to the Celtic Church of Britain or according to that of Rome. It chose the Roman, thereafter allying itself to the power and authority that was the Roman church.

The calendar has long been used as a source of power. The traditional calendar on Simbo in the Solomon Islands is a tally of moons, called the *pepapopu*—a tipplish word, a name to conjure with—for, at least magical, at most divine, calendar-knowledge was considered so powerful that when the tally was first revealed to a mythic calendar-keeper, he was killed by jealous rivals. The human calendar-keeper was thought to control the movement of time itself. In many societies, priestly power stemmed from their timing functions, as in the Nigerian Ibo village described by Chinua Achebe, in *Arrow of God*. These included identifying the moment of the new moon, deciding when the pumpkin festival should take place and announcing the time for the yam harvest.

Achebe describes how, as a result of white colonialism, the traditional god fell silent—as did the priest. With no call to harvest, the people began to starve.

The Romans used the Egyptian "Wandering Calendar" which had 365 days, but which, every four years, had moved a day ahead of the seasons, causing, according to gruff Romans, administrative havoc. Julius Caesar in 46 B.C.E. added an extra day every fourth year to correct it, but *carpe diem* wasn't good enough. Seize a day, seize a month. Caesar month. Seizing symbolic power over time, the Roman senate renamed a month in honor of Julius and then one in honor of Augustus, so July and August have ever since blazed their sunny immortality down a trail of summer years.

But it wasn't quite so easy; with a leap year day, the calendar started losing 0.0078 days a year against the solar year. In March 1582, Pope Gregory XIII sat down and worked out how to rid the calendar of a pesky leap year too many, ensuring no one would need to adjust their set again for 3000 years. He cancelled ten days forthwith, to redress accumulated errors, so the day after the 4th of October would be the 15th. Ticketyboo. And poked his quill back in the ink. Flanders and some parts of Belgium missed Christmas, as they didn't make the change until December 21st, which was then followed by January 1st. But altering the calendar is an act of almighty power. Watch one reaction. The English, more gloriously—and disastrously—emotional than they ever admit, threw a complete tantrum. The Pope, of course, was a *Catholic* and had got his papal paws on the calendar of Reformation England. Protestants protested. The English refused to accept the Gregorian calendar until 1752, when, despite nearly two centuries' notice, they still seemed to come upon it all unawares. Eleven days were disappeared in September 1752 and there were riots, as Hogarth depicted: people thought that they were losing eleven days of their lives and of their wages. In 1873 and 1875, Japan and Egypt adopted the Gregorian calendar, the first almost-universal calendar the world has known.

In the struggle for world domination between Christianity and the Mithraic mysteries, the treatment of Oceanos, the "stream of time," was interesting. Oceanos once had a proud position in iconography, producing Mithras from his waters of time. After Christianity won the battle, the image changed subtly but elo-quently; Oceanos is pictured at the feet of Christ, not giving rise to Christ as he had to Mithras, but subdued by Christ. Time was crushed by Christianity, as it was to be over and over again.

Anno Domini was invented in 525 by Dionysius Exiguous—or Dennis the Small—a small step for Dennis, but a big step for the power-hungry Church, for this, the first slab of global time, was laid in Christian stone. Some five years later came a hugely influential and novel attitude to time, the Rule of Saint Benedict.

Night. Sin-black, sixth-century night was colonized by this *Rule*, into a white order of time, bells rang through the nights and days proclaiming the hours from *Lauds* at roughly three-thirty a.m. to *Compline* at roughly eight p.m. The first "alarm" clocks rang a bell (a cloche, hence the word clock) to get the monks out of bed, and thereafter came centuries of sleep frightened out of sleep's wits. Churches and monasteries began to divide the day and diminish the night, colonizing "*Tempus Nullius.*" From keep-ing watch at night to the watches and clocks so beloved of Chris-tianity: wake, watch and pray were the watchwords of the Church. (In the so-named Apostle clocks, the apostles represented the hours and when twelve o'clock struck, the twelve apostles trooped past the figure of Christ, the "keeper of time.") Chris-tianity sought power over time itself.

There were daily, weekly and yearly timetables for prayer, work and sleep; for bathing, bloodletting and mattress-filling. This coercive time-discipline exerted a power over human nature to prevent it tripping away its days in play: idleness, that impish, happy spirit was decreed "the enemy of the soul." The monasteries represented time moving with the rhythm of a machine. The Christian church continued its taming of time, charting, fencing, bounding its wilderness, into the *Offices* of hours and from the twelfth century onward, churches erected belfries and campaniles. (Christian spiritualists had clockwork visions: in 1334, Heinrich Süse, a German spiritualist, had a vision of Christ as an elaborate clock.)

Public clocks began to appear, and from around 1345 people gradually began to use hours divided into minutes, and minutes divided into seconds. (The treatment of time parallels the treatment of land and it is fascinating that in this period which began measuring the moment, space became tightly measured with the use of perspective in art. Before the fourteenth century, artists painted things as they were known to be over time, not things as they appeared at one single given moment. With perspective, though, the moment becomes fixed.)

Not content with exerting power over its own believers, from the early Middle Ages, missioneering Christianity attempted to impose its rhythm of feast days and work days worldwide, using clockwork as a sign of spiritual potency. In Spain's *Reconquista*, Muslims were forced to celebrate Christian feasts in Castile, resulting in the rebellion of 1264. Matteo Ricci, Jesuit missionary to China in the 1580s, used clocks to gain entrance to the emperor's court, for his calendar was more accurate than theirs and his clockwork remarkable; his "self-ringing bells" fascinated the Chinese and, in this quasi-universal language of time and power, saying "my timekeeping's better than yours" means "my god's more powerful than yours." For this was the period when God-the-Clockmaker was the apex of spiritual imagery and if his work was clockwork then the workings of clocks represented the workings of the very mind of God.

Pre-colonialism, the Tewa peoples of New Mexico had a calendar of rituals carried out during the year, such as the rites of the solstice, for instance, or of "bringing the buds to life." After the Spanish invasion, the Catholic church specifically targeted the calendar as a *locus* of power and suppressed the timing of indigenous religious activities, while promoting the timing of the Christian year, in a pattern which was repeated across the world.

St. Kilda, a hundred-odd miles west of Scotland, has been called "the last and outmaist isle" of Britain. An indigenous culture thrived there for hundreds of years, but it was brought to its knees, literally and figuratively, by the time of the Christian church, most particularly in the grim Sabbatarianism of nineteenth-century missionaries, the Reverend John Mackay above all. Mackay (who owned the only watch on the island) stayed twenty-four years and effectively banned work—the fishing or fowling on which the islanders depended for their very survival—from Saturday to Monday and frequently on Tuesday and Wednesday as well, with devastating results.

When missionaries arrived among the Algonquin peoples of North America, the Algonquin called clock-time "Captain Clock," because it seemed to command every act for the Christians. (In *Gulliver's Travels*, too, the Lilliputians observed that Gulliver's God was his watch.) Addressing a Peruvian congregation, a contemporary Texan missionary declares that a meeting will take place "At ten o'clock, punto, ten" on the dot of white Christian time. Missionaries, past and present, have insisted on dictating the time-use of those they would convert; the schedulization of time in mission schools, the disciplined timetables and new rules of time. Richard Gott, an expert on Jesuit history, says:

Time, a territory once wild, was tamed by Christianity: so people were corralled in, and dragged into a time enclosed and fenced as surely as land in Britain was by the enclosure acts. When the Tikopia people of Polynesia converted to Christianity in 1928, they still lived in their "own" time, celebrating a ritual cycle, a "circulation of sacredness in the course of a round of ceremonies," as anthropologist Raymond Firth describes it. By 1956, the ritual cycle was abandoned and all the circularity, the "cycles," "circulations," "courses," and "rounds," were gone, enclosed by the conquering, linear fences of the more powerful Christian time.

This still happens today. Meeting representatives of the Guaraní-Kaiowa Indians from Mato Grosso do Sul in Brazil, it is clear that their sense of a calendar is significant and sad. As so often, the treatment of land reflects the treatment of time; their land is being stolen by cattle ranchers for burger chains and their calendar stolen by Western missionaries, for the burgerization of their time.

New religions have invaded our villages. It's cultural extermination; they say our rituals are of the devil. Because of them we started to follow the white calendar and now we're sad we changed to the Western calendar because it means we're living in the modern world. Before then a year was a very long time for us.

A gibbet, a drawbridge, flags and turrets and oil drums. Made of scrap metal, wit and anarchy, place of white cider and Attitude.

Welcome to Fort Trollheim, a "Hill Fort" built (and destroyed) in the mid 1990s. Trollheim, on the site of the road protest at Fairmile in Devon, England, was one of the most spectacular creations of the British direct action movement. And it had its manifesto:

> This is the Independent Free State of Trollheim. We declare that as a Free State we have no allegiance to the United Kingdom and its environs, except Essex, Glasgow, Brixton, parts of Hackney and a few small pockets of Devon. We do not recognise the government or monarchy's authority over us, except giros and legal aid. . . . We do not recognise history, patriarchy, matriarchy, politics, communists, fascists or lollipop men/ladies. . . . We have a hierarchy based on dog worship. Our currency is to be based on the quag barter system. *We do not recognise the Gregorian calendar: by doing so this day shall be known as One* . . . Be afraid, be afraid all ye that hear. Respect this State.

To choose your calendar is to choose your politics, for politically active Trolls at Trollheim, just as it was for the French Revolutionaries. The relationship of time with political power is so profound that Lewis Mumford, in *The Myth of the Machine* (1967), writes that the first evidence of fascination with time, through astrology, "coincided with the very birth of kingship."

In C.E. 807, an embassy from Haroun al Rashid to Charlemagne's court took a waterclock, to the murmuring admiration of the courtiers. Medieval dukes and princes flaunted mechanical clocks, expensive and therefore prestigious, but also hinting at that ancient power; those who measure the hour hold sway over time itself.

In 1370, Charles V of France gave an order that all clocks were to be set by the magnificent clock in his palace; he was the ruler of lands and now he would be ruler of time. (This clock was severely damaged in the French Revolution—articulate vandalism this, to attack such a symbol of aristocratic domination.)

The prestige—and power—of early clocks is illustrated in a story from Prague where there is a famous astronomical clock, built in the early fifteenth century. Mechanical figures, including Christ and the apostles, and Death, Greed, Vanity and a Turk, were added later by a brilliant clock-technician, making the clock a wonderful—and unique—creation. So pleased was the Town Council that in appreciation of his work, they ordered that he be blinded so that he could not make a similar clock for anyone else. In retaliation, the blind and furious man broke the clock and it stayed unmended until the mid-sixteenth century.

To teach someone a different clock and calendar is one of the most subtle but most profound aspects of imperialist power—colonialism of the mind. When ancient China had colonized some new region, the phrase they used to describe this act was at once sinister and telling—the people of the new territory had "received the calendar." For Chinese emperors, the Mandate of Heaven involved a "stewardship" of time; the ruler, identified as a paramount sign of the times, was responsible for time. When a new dynasty came to power it altered the calendar of the previous dynasty and the start of a new reign was dated as the first day of the following new year, thus the new emperor ritually regenerated time itself.

Just as the Romans counted the years *ab urbe condita* (from the founding of the city) tying calendar-time to their own prestige: so in the French Revolution, 1792 was designated Year One. Power had changed hands so time must change hands and any association with the church, shaper of the calendar for hundreds of years, was thrown out. Time past was to be guillotined from time present.

There were to be ten-day weeks, each day of ten hours divided into a hundred minutes starting on the first day of the Republican Era: September 22, 1792, the new New Year's Day. "Time," said the reformers, "is opening a new history book." Between 1791 and 1794, festivals were instituted by law. The days were renamed, *Primidi, Duodi*, and so on, and the months were renamed to accord with the seasons; *Vendemiaire* was September, October was *Brumaire*, followed by *Frimaire, Nivose, Pluviose, Ventose, Germinal, Floreal, Prairial, Messidor, Thermidor* and *Fructidor*. (They were translated into English like a parade of the twelve dwarves as Wheezy, Sneezy, Freezy, Slippy, Drippy, Nippy, Showery, Flowery, Bowery, Wheaty, Heaty, Sweety.) Walter Benjamin noted that in the July revolution "the clocks in towers were being fired on simultaneously and independently from several places in Paris"; revolt against the clock.

Zion Ward, the nineteenth-century "messiah," instructed his disciples to count the years from the date of his "illumination"—1826—which he called the "first year, new date." "In the South," said one commentator, writing of the American Civil War, "the war is what A.D. is elsewhere: they date from it." When Mussolini came to power in October 1922, a new calendar was introduced in Italy, marked "Era Fascista." Ezra Pound wrote a calendar in the spring of that same year which supposedly marked the end of the Christian era and the beginning of the "YEAR 1 p.s.U."—in other words Year One, *post scriptum Ulixix*, in homage to Joyce.

Britain "ruled the waves" in the days of empire and through the oceans had power over vast parts of the world. How? Through clocks. In 1714, a petition was offered to parliament, proposing a prize for the solution of the longitude problem. "The discovery of longitude is of such consequence to Great Britain for the improvement of Trade . . . the lasting honor of the British nation is at stake." Through mastering, with chronometers, the mystery of longitude, thus rendering seas navigable, the British paved the pathless oceans and paved them in the pursuit of power, prestige and profit. The chronometer became a tool of political power, a weapon of empire and the handcuffs of slavery.

The most accurate clocks were kept at the Greenwich Royal Observatory, the center of this maritime nation and the center of empire. Reeking with the language of imperialism and smug with the knowledge that time is power, the chief clock at Greenwich in 1852 was called the "master" clock, it sent out signals to "slave" clocks in Greenwich which sent further signals to other "slave" clocks at London Bridge. Today, at Greenwich, there is a plaque (in a dismally unprominent position on a run-down housing estate) which quietly commemorates the slave trade in African people which "was to enrich England for centuries and correspondingly destabilise and impoverish Africa." The same language of clock-power is still used today—there are "Master Clocks" at the Rugby Radio Station for the MSF service of broadcast standard frequency and time. In keeping with both maritime history and the slaving past, in the U.S., the Navy serves as the country's official timekeeper, with the Master Clock facility at the Washington Naval Observatory, while the website of the National Institute of Standards and Technology in Boulder, Colorado, advertises workshops featuring the design of Slave Clocks.

If time at sea represented Britain's power over other nations, within the country, time for railways represented London's dominance over the provinces. Although time varied between the west and the east of the country, train timetables needed a uniform time; so London Time was decreed in 1840 to be that standard. The Great Western Railway printed its timetables accordingly, introducing London Time at its stations. Plymouth and Exeter, in the West, hated this expression of the capital's political dominance and refused to accept it for years. London Time finally became law in 1880. The clock at the Bristol corn exchange has three hands; an hour hand and two minute hands because they

register both Bristol and London time, Bristol eleven minutes behind GMT. (Today, in the former Soviet Union, Ukrainian time
is slightly different from Moscow time, and trains on at least one
line set off at unpredictable times, depending on the nationality of
the driver.)

1883 saw time zones introduced in America. By international
agreement in 1884 in Washington, the world was divided into
twenty-four time zones, beginning with Greenwich Mean Time.
Suitably for a nation of shopkeepers, who had long linked commerce and clocks, the first global export was time itself, GMT.
(The impulse toward globalization began early: the first object
printed on Gutenberg's press in the middle of the fifteenth century was a calendar and not, as is usually thought, a bible.) As this
became the universal time measurement, it signaled the destruction of other ways of counting times and marked the hegemony
of the one, Western—specifically British—way, dictated by British
imperial power, as seventy-five percent of merchant ships already
followed GMT. The French, never slow to see a slur on their national pride, were at the barricades before you could say *Bob est
ton oncle* and, until 1978, they continued in law to call GMT "Paris
Mean Time retarded by nine minutes and twenty-one seconds."
The concepts of imperialism extend to outer space where astronomers use GMT to make their predictions, though they call
it UT—Universal Time.

In the early years of BBC radio, the time signals were announced in voices heavy with *auctoritas*, laden with *gravitas*, declaring, through their ringingly upper-class accents and superior
tones, the age-old equation of time and power—in this case social power. The chimes of Big Ben (built in the 1840s and 50s at
the height of empire) ring out the same message, for while its tolls
dignify and authorize news time on television and radio, with inherent prestige, its geographical position at the Houses of Parliament tells of political power and its striking the hour of London

time out across the world recalls—at a stroke—its imperial past, its trading history, its imports of slaves and its exports of global time itself. The design of the Houses of Parliament was itself a subtle use of time, using the past to authenticate power, for it was built in the Mock Gothic style, designed to look far older than it was, to pretend to a more established and thus more authoritative past than it actually had. As Eric Hobsbawn argues in *The Invention of Tradition*, this phenomenon has been frequently used to invest political structures with a past they do not possess.

Pol Pot declared 1975 Year Zero. Hitler defined his political ambitions through time: the Third Reich was to last a thousand years. Stalin tried to cancel Sunday: between 1929 and 1940, he introduced the five-day then the six-day week. In order to demonstrate its power over the other parts of the Soviet empire, Moscow Time was decreed the standard of communist time, to which everyone must synchronize their time. Synchronization is highly political; totalitarian states adore it, from the vast synchronized gymnastics of fascist countries to the synchronized *Heil Hitler* salute. Synchronization illustrates the totalitarian desire to subsume the individual into the mass. It also, similarly, represents a wish to blur specific, various times into a global monotime.

From the military command "synchronize your watches," to the recent Gulf conflicts, time has been enlisted as a weapon of war and tool to power. The second half of the thirteenth century saw technological advances in the making of mechanical clocks together with the first cannon. In the fifteenth century, soldiers going to war took their own roosters with them to wake them early. Command depends on timing; in the Flanders campaign of 1708, Henry Sully, famous old English watchmaker, was engaged to go to war with Prince Eugene of Savoy and the duke of Aremberg. In the Gulf War, time was used in the split-second bombing of Iraqi targets, and in the instantaneous TV reporting to win the public's political approval. The deliberate time-brinking in the

later Iraq conflict teed up a volatile—and therefore highly manip-
ulable—public mind.

Colonialism has always involved the political manipulation of
time and memory. In Ireland, Eamon de Valera, leader of Sinn
Fein, is said to have remarked after the Easter 1916 uprising
against the British that the difference between the countries was
that "the British never remember their history. And the Irish
never forget." It underscores all colonialism; the vanquishers can
somehow never quite recall the exact path they took to their present dominion, while the vanquished cannot forget their hu-
miliations, deaths and defeats. In Papua New Guinea, the era of
colonialism was called *taim bilong masta* in Pidgin.

Right on the border of Texas and Mexico is a city straddling
two countries. The Mexican half is Juarez, the Texan El Paso. The
Americans of El Paso joke about the other side working "on
Mexican time"—faintly pejorative and definitely patronizing.
The same terms can be used in a deliberately derogatory way:
latter-day colonialists describing people "working on Indian time,"
or "on Maori time," mean a sort of idle and disrespectful unpunctuality. Anthropologist Edward T. Hall writes of the colonizing of the Lakota Indians. They had, he said, no word for "late"
or "waiting." The first thing the superintendent of one reserva-
tion did was "to force them to learn time." Pretending, as the
West routinely does, that its time was *the* time, the one and only
monotime, what the superintendent actually meant was learning
an ugly, shredded time—measured clocked-time—and bowing to
its language of power. Efficiency. Punctuality. Work-discipline.
"Fill the unforgiving minute with sixty seconds' worth of distance run," as Kipling wrote in "If"—the poem most beloved of
British imperialists. "There was nothing noble-minded men
could not do when they discovered they could slap time on their
wrists just like that," pointedly remarks Antiguan·writer Jamaica
Kincaid in *A Small Place*, a powerful critique of colonialism.

Before the Europeans came, the Navajo used sun-time, the hour was the sun. And time was not money. After contact with white time, the Navajo tried to adjust to counting arbitrary, abstract hours and, for the first time, they equated time and money.

The story of the Trobriand Islands' calendar is a nugget of time, power and colonialism. Traditionally, the responsibility for the calendar was divided between four districts. One, Kiriwina, was dominant, milking the calendar to increase its power by choosing when to insert certain intercalary months to ensure that various "harvest exchanges" of yams and other foods would, like pass-the-parcel, just happen to end up in Kiriwina's lap when the calendar music stopped. Enter Leo Austen, administrator of the Trobriand Islands in the 1930s, brimful of work ethic and steaming with industrial capitalism. He came, he saw and, yes, he altered the calendar for, as he noted, spluttering with Protestant indignation, these extra months happened during non-work times of the year; it was an excuse for laziness. So he put the extra months into the work season instead. (His calendar was in turn overruled by the dominant Gregorian calendar.)

In our age, the leaders of the Zapatista peasant uprising in Chiapas, Mexico, insisted *their* time was not the time of the Westernized, modernizing Mexican government. The Zapatista leaders took their orders from the peasants—a process both very slow and completely unschedulable. In their negotiations with the government in the mid 1990s, the leaders commented: "We use time, not the clock. That is what the government doesn't understand." Zapatista spokesperson Subcomandante Marcos, speaking to crowds in Mexico City in March 2001, evoked time in his rich, suggestive way, interweaved with the names of indigenous peoples: "Tenek. We come from very far. Tlahuica. We walk time . . . Zoque. We carry much time in our hands . . . Raramuri. Here the dark light, time and feeling."

In the Pichis Valley in Peru, "development" workers waded

into the territory of the indigenous Ashaninca people in the 1980s. Arriving, full of projects, the workers claimed the Ashaninca didn't "know how to work." The Ashaninca invited them to stay and enjoy themselves with a canoe-load of manioc beer. For their hospitality, they were lectured on laziness. Adventist missionaries had also come to the area in the 1970s. One writes scaldingly of "natives . . . who don't know what work or progress is." The missionaries corralled the people into mission stations to bring "order" into their lives—meaning in effect that the people should bow to Western, Christian time-use and time-values, the women working full-time in the kitchens and the men working nine-to-five, while life on the stations was tightly scheduled with prayers, schooling, work and meals.

Those who have power chain the time of others: never more so than with slavery. Colonial memoirs are full of "lazy" slaves and "idle" laborers, of commands to work "at the double." For the enslaved themselves, concrete slavery has always been marked by a profound desire to seek an abstract freedom and dignity through time, so slaves have used time as protest with go-slows, subverting authority by delay and by slowness, stretching time with silence, shortening it with song, finding a time-freedom in unpredictable presences, unforeseeable absences, momentary disappearances and sudden reappearances.

A Digression.

J.G. Ballard, in *Chronopolis*, describes how an over-clocked world undermines people's dignity and sense of freedom, leading to a revolt against clocks, so that clocks were made illegal and winding up clocks was a "time crime" punishable by imprisonment.

In 1791, Jeremy Bentham wrote *Panopticon*, about a proposed circular form of prison around a central "well" from which warders could observe every inch of the prison, every hour of the day. Surveillance like this is an insidious imprisonment, for it observes not only where someone does something but when they do it, enslaving their integrity of time (that freedom which of all freedoms is the most metaphysical). The prisoner must not only "do time" but must be seen to do it.

An attitude of time-surveillance spreads insidiously today. *Time* magazine recently claimed that the average New Yorker is captured on video cameras and closed-circuit TV up to twenty times a day; their time is watched. Big businesses install hidden cameras to spy on workers, and employees of Olivetti, the computer company, wear electronic tags, enabling bosses to track them to check they are not *wasting time*. It erodes freedom with a slavery of self-observation, a creeping guilt trying to avoid offenses against the clock, a sense of chronocriminality. Twenty-six million American workers are monitored by management to check their efficiency. One clerical worker found her video screen flashing at her "You're not working as fast as the worker next to you!"

The panopticon links surveillance, guilt and "doing time." This matrix is the opposite of innocence, which is guiltless, innocent of time and of surveillance. Picture, for instance, a child's unwatched innocent timelessness. Surveillance spoils this innocence. Noticing a child makes it notice itself, makes it aware of the passing of adult time. To be innocent is to be innocent of clock-time and of watch-observance.

Observatory is a horribly accurate word; observing time is a tool to power from Newton's Observatory to Cook's Observatory in New Zealand, the U.S. Navy's Observatory, Bentham's observation-prison and Gradgrind's Observatory in Dickens's *Hard Times*, a "stern room with a deadly statistical clock in it, which measured every second with a beat like a rap upon a coffin lid." And Observatories are hated—Gradgrind's children hate their father's observatory. In 1894, an anarchist tried to blow up the Greenwich Observatory; it was also the site of anti-millennium protests. The Cook Observatory is constantly defaced by Maori political graffiti.

If ever an age forged the chains linking time and power, if ever an age watched time and enslaved it, it was the Industrial Revolution, an era which altered the experience of time more dramatically than any other. Work, work, work, was the whip word. The watch-chained Victorians shrilly maintained that all hours were to be spent either building economic power or spiritual power, envisaging a system of "spiritual accountancy" whereby hours spent working or praying would accumulate future rewards, whereas playing, lingering, jelly-making, gambling or giggling were coins of hours badly spent—every trifling, whistling one of which would cost you dearly in the afterlife.

John Wesley, Methodist. The very word is like a fetter on time's freedom. (At a Methodist Sunday school in York in 1819, teachers were fined for unpunctuality.) Methodist: austerity such that no hour can slip away in glee nor a body snooze away a fuggy morning. Here he is, Mr. Wesley, in a 1786 tract on *The Duty and Advantage of Early Rising*: "By soaking . . . so long between warm sheets, the flesh is as it were parboiled, and becomes soft and flabby. The nerves, in the mean time, are quite unstrung." One Rev. J. Clayton wrote in a 1755 pamphlet, *Friendly Advice to the Poor*, of "that slothful spending the morning in bed: the necessity of early rising would reduce the poor to a necessity of going to Bed betime; and thereby prevent the Danger of Midnight revels . . ."

The rising middle classes, responsible for colonialism and stiff-upper-aspidistras, capitalism and the loss of play, could do one thing well; they could certainly get out of bed. Getting out of bed betimes, ah me, the hardest thing you have to do all day and you have to do it *first*. But they were not content with their own rising, their own useful productive time. They looked beneath them and saw *the poor*. What did poor people do all day? They slept in, they slept out, they slept around and they slept around the clock. They idled, they hummed, they chatted, they drank. They honored Saint Monday from Tuesday to Sunday, they had holidays all over the time, they ambled and they gambled. The middle classes purpled with a collective Mock Gothic rumble. John Foster writing *An Essay on the Evils of Popular Ignorance* (1820), bemoans the fact that manual workers had, after work "several hours in the day to be spent nearly as they please . . ." *As they please*, if you please, as if their time were actually their own. The presumption!

Though the middle classes pretended that productive time-use was about morality, it was actually about class politics and power, in such an age of power, trade power and colonial power. For the middle classes made their money out of other people's time—it suited the ruling classes very well that those below them should use their time to work to increase the profits and power of middle-class capitalists. As landowners took over the common people's land in the enclosures, so new capitalist timeowners took over the common people's hours in the factory-enclosures of time.

In an extraordinary conflation of ideas, time was changed forever. The economics of capitalism combined with the Protestant work ethic, the sour snobbery of class structure, the moral strictures of Puritanism, the stance of colonialism, new urbanization, Christianity's hell-now, heaven-later accountancy of time, and the rise of synchronized working in factories—making a time of machines. When Lewis Mumford writes in *Technics and Civilization* that "the clock, not the steam engine, is the key machine of the modern age," machine is the word. Most early factory machines were actually made by clockmakers. Machine production in factories required *people's* time to be mechanized, to be as clockwork. This era was driven by new types of power: steam-power, water-power, coal-power and now—added to these—there was clock-power.

And all these were underpinned with the theories of science—Newton above all (a misanthropic alchemist from Grantham) who had offered the intellectual justification for mechanical time which so despised human time, Newton who had said that time was absolute and uniform, factory-time in overalls. Time was Granthamized. The Victorians loved him, inaugurating, at Saint Peter's Hill, Grantham, in 1858, a statue of him, two tons of Granthamite misery for two centuries of hard time.

Just four years before, in 1854, *Hard Times* was published—a fabulous, blistering portrait of factory-time, describing Coketown (a New-town place for Newton times), where "every day was the same as yesterday and tomorrow, and every year the counterpart of the last and the next." "Time went on in Coketown like its own machinery: so much material wrought up, so much fuel consumed, so many powers worn out, so much money made."

What was lost? Time varied, elastic and colored. Time local, mischievous and ribboned. Time seasonal, haphazard, red-lettered and unpredictable was gone. Time was wearily dreared out of bed. Convenienced and colonized. Mapped. Leveled. Privatized. Enclosed. Counted in and accounted out. Factored in and factoried out. Days of equal width, hours of equal length of watched time replaced the stretchy hours of sun-time and the seasonal days of times gone by.

Factory-discipline and clock-power needed predictable, standard days of work. Gone was the honoring of Saint Monday—which cobblers, colliers, cutlers and cockneys, potters, printers and picnickers worshipped. Here, France and England could agree, as Duveau, the French historian wrote in 1946, "*Le dimanche est le jour de la famille, le lundi celui de l'amitié.*" (Sunday is the day for the family, Monday the day for friendship.) Saint

Monday, in effect the patron saint of hangovers, was reviled in Victorian temperance tracts.

The time of festivities, too, was a focus for power struggles for they often dramatized local class politics; the right of the parish poor to collect certain doles, or gather wood, the legitimization of grazing or gleaning rights. Many such festivals were outlawed by the Victorian middle classes; the workers' holidays overruled. It was of course under Margaret Thatcher (a misanthropic chemist from Grantham), very much heir to Newton's factory time, and a banger-on about middle-class Victorian values, that May Day, the traditional workers' holiday, was threatened with abolition.

Hard Times has its Granthamite Gradgrinds and Bounderbys, the factory-owners and politicians seizing power over the grim lives of the factory workforce, but there is more; a third class, the "circus people" who represent pre-industrial time, the jumbly, curly time which was lost, the people who drink far too much and have far too much fun and refuse to believe the stricture that time is money so Mr. Bounderby of Coketown remarks severely to the circus people: "We are the kind of people who know the value of time, and you are the kind of people who don't." Coketown's time is the "hard" time of the title; the circus people represent "soft" time; humane, funny, whimsical and irregular, a time for children and animals. Time is always a genderizable concept and, in this instance, the differences are stark. "Circus" time, round like the circus tent or a circle is female time, represented in the book by a circus girl, Sissy, sisterly in nature and name. (Her second name is "Jupe" the French for a woman's skirt.) "Hard time" is represented only by men, forcing time straight like a ruler; in that world, Gradgrind will not call her Sissy Jupe but insists she is "Girl number twenty"; the sequence of numbers represents the linearity of time and an interchangeable child for interchangeable, mechanical times.

Punctuality and regularity in the work force became, with watches, first a possibility and then a necessity. It wasn't that people didn't work before the Industrial Revolution but, as E.P. Thompson says in his funny and brilliant essay *Time and Work Discipline*, the work pattern had been "one of alternate bouts of intense labor and idleness whenever men were in control of their own working lives," but in this period, power over people's time was taken away from them and put under the control of factory owners.

Josiah Wedgwood introduced the system of clocking-in; and Crowley went one-up, writing *The Law Book of the Crowley Iron Works*, a penal hundred-thousand-word-thunder-code; time-sheets, time-keepers, time-informers and time-fines: his employees' hours of service were calculated "after all deductions for being at taverns, alehouses, coffee houses, breakfast, dinner, playing, sleeping, smoking, singing, reading of news history, quarrelling, contention, disputes or anything forreign to my business, any way loytering."

This power over working people's time was deeply resented. A Scottish industrialist spoke of "the utmost distaste on the part of the men to any regular hours or regular habits" and workers' shock that "they could not go in and out as they pleased, and have what holidays they pleased." One Yorkshireman wrote in 1830 that the factory system was "a state of slavery more horrid than . . . that hellish system 'Colonial Slavery.'" A nineteenth-century Dundee textile worker commented on the abuses of time-power thus:

> In reality, there were no regular hours . . . The clocks in the factories were put forwards and backwards, morn and night. Instead of being instruments for the measurement of time, they were used as cloaks for cheatery and oppression. A workman was afraid to carry a watch, as it was no uncommon event to dismiss

anyone who presumed to know too much about the science of horology.

The class-based character of watch-owning was seen when in 1797–98, the government tried to impose a tax on clocks and watches which failed because it was viewed as an attack on the middle classes.

E.P. Thompson quotes a mill worker in 1857, his words illustrating how time-keeping was a symbol of power:

> We worked as long as we could see in summer time, and I could not say at what hour it was that we stopped. There was nobody but the master and the master's son who had a watch, and we did not know the time. There was one man who had a watch . . . It was taken from him and given into the master's custody because he had told the men the time of day.

In America, the earliest textile factory complex was at Lowell, Massachusetts, which opened in 1823. It had a surveillance system and a highly regulated labor force, its rhythms governed by the clock and machine. Bells in the cupolas rang insistently, tolling out the lie that time is money; bells tolled the workers awake, to their jobs, to and from meals, curfew and bed.

With the industrial revolution and factory-employment, people ceased to own their *own* time; the workforce had to demarcate time which "belonged" to an employer. Watch-ownership increased, for not only were they prestigious, they were also signs that though you may not own your own time, you could own the measure of it. (Interestingly, today, the refusal to wear a watch is a subtle status-signal: "My time is owned by no one. I am no one's slave.") The first joy of watches—and their inherent prestige—has rarely been so deliciously expressed as by a Sussex diarist in 1688 who notes that he "bought . . . a silver-cased watch, which cost me 31 i . . . This watch shewes ye hour of ye day, ye month of ye year, ye age of ye moon, and ye ebbing and flowing of ye water; and will goe thirty hours with one winding up." What he bought, of course, was ye status and ye pride, ye innocent pomp as ye plump round prose shewes so gleefully after all ye years.

No such power over people's time could have been usurped without protest. With eloquent violence, workers in Britain, particularly in Lancashire and other textile districts, during the 1820s and 30s, smashed the clocks above the factory gates, the loathed symbol of a new world order which had stolen the power over their time. Trade unions—very quickly made illegal—took on first the abuse of time so, seeking shorter hours, the early unions threatened go-slows until they won the ten-hour act of 1847. The 1848 revolution was partly about time; a desire for the "three eights"—eight hours' work, eight hours' sleep and eight hours' play. Karl Marx highlighted time's role in capitalism, arguing that the regulation and exploitation of workers' time was a key aspect of capitalist philosophy and warned of the "overconsumption" of workers' time (a prescient phrase for today's "greedy institutions"). In Samuel Butler's satirical novel *Erewhon*, the workers destroy their machines to stop the machines destroying them. It includes an incident concerning the carrying of a watch, and there is a museum for bygone evils, considered the right place for watches. In Paris, May 1968, student protesters stopped a clock at the Sorbonne, writing below it *Nous y mettrons le temps*, meaning both "Let's take time off" and "We will decide what the time is."

The *order* of time runs through much of the history of time and power. To the clock has been ascribed the ability to order life; religious *orders*, of course, first ordered the disorder of time, their bells "lifted all things unto a sphere of order," says clock historian Carlo Cipolla. In 1481, a petition to the Town Council of Lyon asking for the town to have a great clock, says: "If such a clock were to be made, the citizens . . . would live a more *orderly* life." Richard Palmer of Wokingham, England, in 1664, giving lands in trust for a sexton to ring bells at 8 p.m. and 4 a.m. did so to "order" the life of the locale. (In *Eros and Civilization*, Herbert Marcuse writes of "the alliance between time and the order of repression.")

And whose natural state was iridescent *dis*order? Who were even more unpunctual than the poor? Who were by nature living in a state of such disgraceful enchantment that they thought the hour of now the only possible time? Who—unforgivably—insisted on seeing the purpose of life to be not work but play?

Children.

The middle class adults saw the power and the profit to be squeezed from childhood time, down mines, and up chimneys. They also may have seen that childhood was a huge conceptual threat to their hard times; childhood is fundamentally in opposition to the schedulization, punctuality and uniformity of factory-time. Children, unstoppably, adhere to circus-time, they have an affinity with Dickens's circus people; it is adults—as Bounderby nearly said—who are the kind of people who know the value of time, children are the kind who don't.

Cometh the hour, cometh the grown-up. Thomas Wedgwood, son of Josiah, designed, as E.P. Thompson writes, "a plan for taking the time and work discipline of Etruria into the very workshops of the child's formative consciousness." He intended it to be a rigorous education system to schedule, coerce and control the wispish free spirits, to form them into mini-machines for factory work out of whose time more Wedgwood money might be minted. Wordsworth—of all people—was proposed as superintendent of Wedgwood's education system. He wrote his furious refusal in the passionate poetry of Book Five of *The Prelude* (1805), spitting about:

The guides, the wardens of our faculties
And stewards of our labor, watchful men
And skilful in the usury of time,
Sages, who in their prescience would controul
All accidents, and to the very road
Which they have fashion'd would confine us down,
Like engines . . .

The diction has it: the "watchful men" observing, from all their observatories, "skilful in the usury of time," are with us still, making machines of children, confining them. Thomas Gradgrind, industrialist of Coketown, a fictional Thomas Wedgwood, has a school, a model of a mini-factory to fill each child with fact, to measure each hour with industry. A school in which bullies and bankers flourish and childhood withers. When his young daughter, "his own metallurgical Louisa," is found delightedly peeking at the circus which has come to town, he furiously removes her with the sneer, "You are childish." And when she grows up, Louisa's most aching reproach to her father is that his theories of upbringing cost her the experience of childhood.

So began a system which reaved the child of its hour, which crushed the dandelion clock underfoot in a march to colonize the wild time of childhood, a march of usurpation which continues all the more furiously today, from toy shops called Early Learning Centers—lest the child be *late*—to "hothousing" children, greenhouse gassing them with a stress which can kill them.

It is perhaps because children are born with such a powerful sense of the now, of wild time, of eternity, that adults need to tame it so ferociously, to spend so much effort teaching children *the* time—or grown-ups' definitions of it. To tell the time, to name the weekdays and the months, to know timetables, to recognize playtime and schooltime and to think up reasons why there should be a difference between them. Children learn a hierarchy of time, who they must wait for and who they cannot interrupt. To obey the clock, pip pip. And as surely as the Bunsen burner evaporates water, the clock causes the evaporation of childhood.

Today, children as young as two in Tokyo attend crammers, as the competitive pace of education demands. Modernity seems to steal time from children; the U.K. press reports that children are becoming more shy because parents spend an average of just eight minutes a day talking with them (quoted in the *Coventry Tele-graph*), while other statistics show that twelve percent of the population spend more time in their cars than with their families (quoted in the *Guardian*). A national survey in Japan found that fathers in single-earner households spent an average of three minutes per day on "family work" including feeding, bathing and playing with their children. (The figure reached nineteen minutes per day at weekends.) In a Management and Coordination Agency survey, 1986, 37.4% of Japanese children "never" interacted with their fathers, for example, eating meals together or talking, on weekdays, compared to 14.7% of American children and 19.5% of German children. 17.1% of Japanese fathers "never" spent time with their children at weekends, and the data seemed similar to U.S. data of the 1960s and 1970s. American fathers spent about two minutes per day with their children in 1965 and about seventeen minutes with them ten years later. Another, conflicting, survey from the University of Maryland, suggested that parents in 1990 spent forty percent less time with their children than they did in 1965.

Children are taught to fit into adult time, though a child's hour lasts far longer than one of ours. This difference is partly to do with the relative time of one's lifespan; as a proportion of a four-year-old's life, an hour is a long time; as a proportion of, say, an eighty-year-old's life, an hour is but a breath. Part of the difference is in the way the brain processes information about time. Part is physiological; children move quickly, their metabolism faster than adults—the river of time therefore seems to run slower than the child whereas it runs far faster than the eighty-year-old. Apart from a difference of duration, there is a difference of direction: old people live in their memories, facing the past. Young people live in their hopes and they face the future.

One of the most striking protests against modernity's brutal and coercive accountancy-time comes in *Momo*, a book for children—though one which is a privilege for adults to read. Written by the German surrealist, Michael Ende, in 1973, it tells the story of Momo, a small girl. ("How old are you?" she is asked. She thinks a moment, then hesitantly replies "a hundred and two?"— for, crucially, no one has ever taught her to count.) Momo is a waifish pixie who turns up in a town to the delight of everyone she meets; shabby and poor, she has one form of wealth—time. She uses her time to listen to people and they love her for it; listening, of course, creates time for someone else, while speaking takes it from another.

The town she arrives in is one where people matter, the barber talks to his customers, the local tavern, Nino's inn, is a social place for old men to while away idle afternoons over one warm glass of wine. But then into this world, stealthily, coldly, come freezing cold men in gray, from the "Timesaving Bank," who count up time and persuade people to save time in order to make money—these gray men effectively steal people's time for "the more time people saved, the less they had." Nino's inn becomes a *fast* food café, McNino's by any other name, and people become obsolete. It is only a child—and one who can't count—who cannot be cheated by the equation time is money. Momo alone can save her town, with the help of Professor Hora who shows her the "time flowers" and remarks: "Time is life itself and life resides in the human heart."

Not a lot of schoolkids know this: the Latin word *ludus* means both "game" and "school," the word school comes from the Greek *skhole* which originally meant "leisure." It was the free time, the playtime, the wild time unaccounted for by demands of state or duties of religion, the time in which a person could "idle well" as Aristotle said, exploring the unmeasurable, the pleasurable, treasures of mind out of time. Ask any playground-philosopher today what school means and you will hear the exact opposite. Measured, unpleasurable, untreasured time marked by school clocks. (And a place where bullies and bankers still flourish.)

In the U.K., early in 1997, the government proposed new ways of scheduling children's time, including the possibilities of open-all-hours classrooms, teaching for babies and summer schools. In one part of the country, a fifty percent increase in the length of the school day is planned and, in another, weekend schooling proposed. Is no time free? Is there no time for children to *be?* To guess, to play, to stare, to count clouds, to catch peanuts in their mouths, to see whose knees are knobblier, to hop, to flick things, to see who can get their tongue furthest up one nostril? "There isn't enough time to *wobble*," wailed a distressed five-year-old to me in the park just now.

But the spirit of childhood will always end up thumbing its nose at adults who try to squash it; the circus classes will always be with us, because children are always with us, living, as they do, pre-industrially, in *tutti-frutti* time, roundabout time, playtime, dandelion-time. Theirs is an idea of time which has no sympathy whatsoever with the Industrial Revolution in all its historic and contemporary manifestations, for children, hardy perennials, are staunch defenders of something far more valuable and much more fun, the time of ludic revolution.

"Good morning," said the little prince.

"Good morning," said the merchant.

This was the merchant who sold pills that had been invented to quench thirst. You need only swallow one pill a week, and you would feel no need of anything to drink.

"Why are you selling those?" asked the little prince.

"Because they save a tremendous amount of time," said the merchant. "Computations have been made by experts. With these pills, you save fifty-three minutes in every week."

"And what do I do with those fifty-three minutes?"

"Anything you like."

"As for me," said the little prince to himself. "If I had fifty-three minutes to spend as I liked, I should walk at my leisure toward a spring of fresh water."

—ANTOINE DE SAINT-EXUPÉRY, The Little Prince

Today, a so-called social revolution is happening to time, the "twenty-four-hour society," in which people work, shop, sleep and play without the "traditional" rhythm of day and night. A recent Future Foundation report showed how this twenty-four hour society will demand twenty-four hour access to doctors, bookshops, banks and schools—all in order to make more money out of time. L. Michael Hager, in *The Futurist* magazine, argues for this twenty-four hour society; treating time "as an economic resource," he proposes the time-sharing of homes and even beds. Like hot-desking, where employees on different shifts use the same office, desk and computer, so hot-bedding—not his terminology—involves "the poor" taking turns to use a bed. He paints a picture of Mexico City where "Anna Gomez"—a generic poor person—an assembly-line worker in a tortilla factory, begins work at 6 P.M. and her children start a typical "day" at school. "Meanwhile, at the family apartment, Anna's parents and mother-in-law recline in beds just vacated by the younger family members." They wake at 2 A.M. and Anna's brother and his wife get into the bed. When they wake at 10 A.M., it's Anna's turn again.

Somehow it is hard to see L. Michael Hager (director of a law institute, with iron-gray hair and a thinker's parting) hot-bedding with his in-laws, but then he is not "the poor." What he misses is that many poor people already bed-share, because they have to—and they hate it. What he also misses is that the human body has intricate inner clocks, evolved over thousands of years into night-sleeping and day-waking. Disturb these and you predispose the body to ill health and to unhappiness, for it is poignantly sensitive to time, certain illnesses hitting at certain hours of the day; heart attacks and strokes at around nine a.m, and asthma at around midnight, while other illnesses are seasonal, from SAD to spring fever. The body is affected by moon times which intrigue the womb and the mind, by sun times which affect the eyes and skin. Even the backs of the knees are light-sensitive, and thus time-sensitive, and their photosensitivity can be used to treat jet lag. A recent study shows that the brain shrinks as a result of repeated jet lag, going to show that travel doesn't, after all, broaden the mind.

Advocates of the twenty-four hour society are inheritors of Franklin's iron, invariant, mechanical thinking, assuming that all hours are all the same, forged in a factory production line and all alike can be used to make money, interchangeable hours for a coin exchange. They are heirs, too, of the factory owners of the Victorian era, the time-tyrants, repeating "time is money" without somehow ever quite answering the question *whose* money is made from *whose* time? (Charlie Chaplin could have told them;

for in *Modern Times*, the factory worker's enemy is the clock. "Section Five. Speed 'er up . . . Section Five. More speed . . ." until he breaks down with pressure.) Fittingly, "Anna Gomez" (herself a factory-prototype) is a factory worker, her time turned into an assembly-line of hours for someone else's money and profit.

The association of money and time has a long history. Counting the hours and counting money irresistibly rhyme in the mind of humanity. In Imperial Rome, Saturn (linked with Chronos or Time) was also called "Moneybags" and "God of the Mint" because the Roman treasury was attached to his temple. Time is Money.

In the third century, Tertullian, church father, pagan-hater and early puritan, whose very name—Quintus Septimius Tertullianus—is a calculation, counted hours, advocating the punctuality of prayers at the third, sixth and ninth hours, saying that these were internationally recognized as "they serve to fix the times of business." The church's "offices" of hours (time) were linked to the offices of business (money.)

In the thirteenth century, clocks became increasingly important for commerce. The urban commercial population often learned to count money by counting the bells of clocks and then used that arithmetic skill in shops and early businesses. Feudalism conceptually linked money with land and livestock, as language remembers, so the *stock* market comes from *livestock*, and an annual *yield* originally meant a crop yield. The value of land rests in its immovability, its constancy; and as the land was a steady plenty of eternal cycles, so time was considered plentiful in those same everlasting rounds. But the rise of capitalism linked money to *time* and, in contrast to land, money's value comes from its circulation, not its hoarding. The value of money—currency—

depends on its *movement*: the word currency comes from the Latin *currere*—to run.

But what runs can run out and become scarce and so both money and, crucially, time became potential scarcities. G.J. Whitrow, in *What Is Time?* says that with urbanization and early capitalism:

> The tempo of life was increased, and time was now regarded as something valuable that was felt to be slipping away continually: after the fourteenth century, public clocks in Italian cities struck all twenty-four hours of the day. Men were beginning to believe that "time is money" and that one must try to use it economically.

There was a shift from what historian Jacques LeGoff calls *"temps de l'Église"* to *"temps du marchand."* The 1481 petitioners of the Town Council of Lyon, who, as the previous chapter noted, asked for the town to have a great clock, offered as one of their reasons that simply having such a clock would bring more merchants to the town.

From the fourteenth century onwards, time started to move faster, as money was linked to time and, further, both were associated with the rise of *urban* economies. (Modern experience reflects this; time not only runs "faster" in urban centers than in the countryside, but the "fastest" places of all are the economic centers right at the heart of cities; the stock markets of Tokyo, London or New York where clock-time rules and a split-second's delay in buying or selling can lose or gain hundreds of thousands of dollars; time is money.)

With early capitalism, time, like everything else, was put to work to build up the capitalist stock. Money could be made out of time. The idea actually began with the Schoolmen of the Middle Ages, for it was they, theologians not economists, who came up with the theory of capitalist accumulation: the "*Treasury* of Salvation," the piling up of merit, through hard work and sacrifice, in order to gain a fabulous future reward in heaven. (And with that whisper "Treasury" Benjamin Franklin is scraping at the door.) The Puritan R. Baxter, in *A Christian Directory* (London, 1673), offered his pennyworth—time is currency, so "use every minute of it as a most precious thing, and spend it wholly in the way of duty." (In 1699, Newton, thought to have mastered the science of time, was made Master of the Mint—the metaphorical master of time was the literal master of money.) Counting time and counting money are bound together as all concepts of saving, credit and debt are based on having a concept of a countable future.

The Protestant church fostered the idea and it is wholly appropriate that it was *Calvin's* Geneva which became the center of watchmaking and of clocks, city of banking and of lending (using time to make money), Geneva, where by history and by mutual mean consent, time was money and money was time. There is a strong link between Protestantism and economic development; in Switzerland, for example, Protestant cantons were centers of industry while the Catholic ones remained agricultural, and a key reason for this difference was the Protestant obsession with the management of time.

In the Industrial Revolution, the equation of time and money was given spiritual backing by the Protestant church. The church was once the "bride of Christ," but in the Industrial Revolution, she got a quickie divorce and married the door-to-door salesman. "Puritanism, in its marriage of convenience with industrial capitalism, was the agent which converted men to new valuations of time; which taught children even in their infancy to improve each shining hour and which saturated men's minds with the equation, time is money," as E.P. Thompson says. People quoting camels and needles were not given much of a welcome.

You hear the rhyme whispered up through the ages, but then in the eighteenth century, enter Benjamin Franklin, American statesman and author, whose very name is a bitten bit of a coin, a little franc, who came right out and said it in 1748; "*Time is Money*." And a new metallic dawn was minted in the coin of time.

In 1751, he says: "Remember that Time is Money . . . he that idly loses five shillings worth of time, loses five shillings, and might as prudently throw five shillings into the sea." (The sea, of course, to Franklin and to his successors, is a waste disposal unit. To sages past and present, the sea symbolizes time and eternity.) Franklin again: "Since our Time is reduced to a Standard, and the Bullion of the Day minted out into Hours, the Industrious know how to employ every Piece of Time to a real Advantage in their different Professions: And he that is prodigal of his Hours is, in effect a Squanderer of Money . . . *Time is Money*."

John Ruskin, author of *Time and Tide*, who also wrote "Time," a poem in blank verse, when he was seven, was infuriated by this brassy meanness and said, aghast: " 'Time is Money,' the words tingle in my ears so that I can't go on writing. Is it nothing better, then? If we could thoroughly understand that time was—itself . . ."

But Franklin had done it, had unleashed the scrupling scrooges of his age and ours, the accountants of time, the standardizers of time to a mercantile percentile, rigid, strict and uniform as coinage itself, Newtonianly absolute. His vicious little lie has swindled the world ever since so you can now buy time, steal it, spend it and save it, you can even "buy" your own "free" time. The success of the deceit is most clearly registered in language for what began as a metaphor has become invisible, pretending to the transparency of truth: time is money. (A now-obsolete meaning of the word minute was actually a small coin: in 1382, Wycliffe recorded that two minutes was a farthing; Benjamin Franklin eat your heartless heart out.)

Columbus, on first meeting the Tainos (San Salvador) was convinced the people should be "*made to work,* sow and do all that is necessary and to adopt our ways . . ." (My italics.) Colonists enslaved people in the "work camps" of the rubber barons—concentration camps by any other name—making their money out of the very lifetimes of indigenous peoples. Few ever survived. Potosí, the "mountain of silver" in Bolivia, was "discovered" by the Spanish in 1545. In total, some eight million people labored and died for the leisure and wealth of white people and to literally finance the capitalism under which they would suffer for the next five hundred years.

"They taught us to work," says Arakmbut Tarzan, speaking of the arrival of missionaries in his lifetime. "Before, we didn't work, we hunted and fished—and had three or four wives—but we didn't work for someone else, for money. From the missionaries, we learned Money, and Spanish and Work. We had to work for money for needs which we didn't have before." Matteo Jicca, also Arakmbut, always associates money with work, and both are foreign concepts. He says that when clocks came, they "marked the times of work, making people the slaves of time. People working in towns and cities have two chiefs, one is your chief at work, the other is the hour, *Jefe Ora.* It creates a lot of pressure to do things punctually, on the hour. Here in the forests, we clearly have responsibilities to fulfill, but if we can't do things, if we can't finish something, then *mañana* is sufficient. Here we live by the rhythm of nature, another world, another life. In the other world, in your world, you can't even stop to take a pee or you lose your job. For us, we work by the climate and by what we want to do."

At Llanes, a town in Asturias, northern Spain, a cigarette machine in a cafe advertises itself, bizarrely enough for a machine selling Spanish cigarettes, with English pound coins stamped with clock faces: time is money. *Hora inglesa es oro inglés.*

In prostitution alone (apart from Spanish cigarette machines) is the phrase "time is money" almost true.

A digression. In *Tristram Shandy*, Laurence Sterne wrote that Tristram's father, a regular man, was most regular in his winding up of the clock, which he did on the first Sunday night of every month. And he "had likewise gradually brought some other little family concernments to the same period," which coy phrase illustrated how Tristram could date his conception. With the uproarious popularity of Sterne's novel, "winding up a clock" became a byword for sex. Clockmakers hated it. There was published *The Clockmaker's Outcry* against Sterne, for, said the anonymous clockmaker,

> The directions I had for making several clocks for the country are countermanded, because no modest lady now dares to mention a word about winding up a clock, without exposing herself to the sly leers and jokes of the family . . . Nay, the common expression of street-walkers is, "Sir, will you have your clock wound up?"

In November 1994, Christopher Bryant, a solicitor from Birmingham, U.K., suffering from overwork, took a rope and hanged himself.

While many tribal societies only work, we are told, for about four hours a day and some do not even have a designated word for work, in the West, where time is money, workaholism bites. Most Americans work longer hours, with more stress, than did their parents' generation. According to the International Labor Organization, Americans work longer hours than people in any other nation. With frequent twelve-hour days and sixty-hour weeks, "presenteeism" entered the cultural lexicon recently; the opposite of absenteeism, its sufferers are neurotically unable to leave the office. One U.K. newspaper, *Sunday Business*, recently promoted itself with the line: "Sunday, your day of rest. Precious time to forget about business, politics, economics, market forces, boardroom battles. Who needs them? Forecasts, analysis, informed opinion. Let them wait. And if you believe that, you're already a day behind. Now Sunday means *Sunday Business*." The paper has a slogan that Sunday is the "first day of the working week." Its effect is horrible; no day is free, there is no time off, all days can be clawed into and subjected to the same gray executive stress. Juliet Schor, in her best-selling book *The Overworked American*, pointed out that Americans today work far longer hours than in the 1940s. Blaming "unfettered capitalism," she lists the consequences: unhappiness, divorce, excessive consumption, *both* overemployment and underemployment and (like Christopher Bryant) suicide.

Believing that time is money has little grandeur. The chair of a large pharmaceutical company defines work-productivity as "half as many people paid twice as well and producing three times as much equals Profit and Productivity." Half $\times 2 \times 3 = $ P. (*Ecce Homo*, eh?) It is a belief that increases the gap between rich and poor, those working at maximum productivity paying others to cook, do child care and even wait in line for them. Schor comments, "the commensurability of time and money . . . transforms a resource that is equally distributed (time) into one that is distinctly unequal (money)."

Akio Morita, chairman of Sony, calculated that in 1989 the average Japanese worked 2,159 hours while the German average was 1,546. Long hours, the lack of secure sabbatical rest and the compulsory spending of free time on corporate wining and dining, has resulted in what Lewis Coser called "greedy institutions" (in his book of that name) which "swallow" their employees' time. In Japan, where these "timebinding" work practices conjoin with rampant overconsumerism, suicides because of overwork are not uncommon, called *karo-jisatsu*. The prevalence of work-related ill health has led to the Japanese word *karoushi* or *karoshi*, meaning death from overwork, entering the English language. In America, in 1999, one man in ten was working at 3 A.M. Largely as a result of overwork, the average sleep time in the U.S. dropped twenty percent during the twentieth century. According to a study by the National Sleep Foundation, reported in the *International Herald Tribune*, fifty-two percent of those surveyed said they spent less time having sex and having fun than they did five years ago. Life's too $hor . . . In the U.K., fifty-seven percent of people are likely to work on Saturdays, and thirty-seven percent on Sundays, resulting in what some commentators have called the "lost weekend," with devastating results for relationships, friendships and family life. Americans spend an average of six hours a week shopping and forty minutes a week playing with their children, according to *All-Consuming Passion: Waking Up from the American Dream*. Other social analysts have suggested that the twin evils of over- and underemployment divided people into two groups; the one "time-rich but work-poor," the other "work-rich but time-poor." Uruguayan historian Eduardo Galeano comments: "Consumer lifestyles make time a scarce and expensive resource: time is sold, rented, invested. But who is the owner of time? Machines. Machines, created to save time or to pass the time, seize control of time." Television has seized control of people's time all over the world: in the States, watching television consumes forty percent of the average person's free time, according to John P. Robinson's *I Love My TV*. One survey shows that the preschool child watches more than twenty-seven hours of television every week. (I can't help almost admiring the French in this regard, for they actually spend more time watching television than working.)

Studies at Cornell University's School of Industrial and Labor Relations found that those who work more than fifty hours per week are more likely to experience "severe" work-versus-family conflicts, and that rates of alcohol use, stress and absenteeism rise for those pressured into overtime. There are about nine hundred books in print in the U.S. on stress and time management. Time-issues are frequent causes of work-stress; long hours, compulsory overtime and, crucially, being unable to control the speed of one's own work.

Marshall Sahlins, Richard Lee and others have suggested that hunter-gatherers represent "the original affluent society" in that a minimum of work produced a maximum of leisure, pleasure and food. Today, in the West, a weird anhedonia co-exists with over-consumption, in that while people possess enormous quantities of property, the time, energy and happiness needed to actually enjoy it is in scarce supply. The Inuit a generation ago said of themselves that they were "rich in knowledge, meat and time." (And, as Hugh Brody comments in *The Other Side of Eden,* were generous with all.)

André Gorz, in *Farewell to the Working Class* writes of the liberation of time and the abolition of work; the slogan "Work Less, Live More!" grew so popular for it chimed with a deep need. He also identifies a reason why employers prefer to reduce the number of workers by ten percent rather than by cutting working hours: "Instinctively they perceive the freeing of time as a threat to the established order. For unemployment is a disciplinary force. When jobs are scarce they can be kept for the most hard-working and submissive." He also notes in *Critique of Economic Reason* how the very notion of "unemployment" was invented "expressly to combat the practice of discontinuous work and to eliminate those intermittent workers who often preferred to lose wages to gain independence from the employer." It abolished a particular—and particularly cherished—time-freedom.

Intermittent hard work followed by indulgent leisure seems to be a human preference, in many parts of the world, at many times. E.P. Thompson considered it to be so in Europe before the Industrial Revolution. Writing of the Native Americans in the Great Plains of Canada, Irene M. Spry comments on the "glory and excitement of the old way of life" when "periods of strenuous activity were followed by periods of leisure, which allowed ample time for ceremonials and the vision quest, sociability, dancing, music, games and creative artistry . . . there was no soul-destroying daily routine drudgery . . . Getting a living was a zestful, community affair . . . it was a sad transition to a sedentary life on reserves, scraping a meager living by cultivation and stock raising under the control of an Indian agent."

Interestingly, there is a reaction to the overwork/overconsume cycle in America—the downshifters, comprising five to ten percent of the workforce and alarming the *Wall Street Journal* which bemoaned that it could cause a dip in the economy. While downshifters in the U.S. and elsewhere refuse to acknowledge time is money, they do see that time is value. (If no one wants to spend their time with you, you feel unvalued. If you are unwillingly unemployed, your time has no value.) Another protest has come from religious groups: Rabbi Arthur Waskow of the Shalom Center, Philadelphia, comments on the importance of the sabbath in the Jewish tradition: "We are taught not only the seventh-day Shabbat but also yearlong observances every seventh year, when the earth rests along with human workers, all debts are annulled and there is free time for celebration and reflection. And in the deepest version of Shabbat, once every fifty years the Jubilee was to bring the re-distribution of all land." *The Nation* quoted Ralph Nader as saying that today it's harder to be a citizen because people are working 160 hours more each year than they did twenty years ago. Nader "gets the most rapt attention from his middle- and working-class audiences when he talks about their shrinking leisure time."

A brief digression. Overworked? Not enough time for yourself? Stressed? Hurried? Always behind? Always letting people down or cancelling things because you can't achieve all you wanted? Everyone has 168 hours every week, yet some will always feel over-pressured, coerced, struggling against time, and others will always get things done calmly, in time, *with* time. So, crucially, how do you change from being one of the first to one of the second?

The late twentieth century, which has spawned such unkind attitudes to work, has also come up with some remedies—Personal Time Managers—purveyors of sage advice. Some of their best is the simplest; don't aim to do so much that you constantly castigate yourself for failure. Enjoy what you have done, rather than punish yourself for what you haven't. Things that can trip you up include a lack of sleep, disorganized work places and an inability to focus on one thing at a time, as well as, obviously, a workload simply too great for anyone to handle.

Time managers suggest writing achievable lists of things-to-do, with very big items broken down into smaller pieces which can be done in a few hours. Don't, don't, don't schedule all your time as gray, gray work-time. Schedule fun. Plan exercise. Give yourself time for pleasure and for ordinary livingy things. Rank your to-do list by importance, 1,2,3, says one time manager. And then, he adds, every now and then just cross off all the threes. Don't promise to do anything unless you can really do it, because this hurts your promisee and snags you up in guilt. Be strategic in planning work, make goals, but be prepared to vary them. Accept the backlog of things-which-have-not-been-done. It's normal. Don't let it be a source of stress, but an indicator of whether you're taking on too much. Have just a few hours every day which are inviolate—no phone, no fax, no visits—and steam ahead with heavily concentrated work then. Be aware of people who erode your time—and don't let them. Schedule working hours; not all hours of the day and night are interchangeable, and what one can do in one hour in the morning might take four hours at night.

In short, be nice to yourself, and remember what the anonymous author of *The Cloud of Unknowing* (1370) wrote with limpid simplicity: "There is nothing more precious than time. Time is made for man, not man for time." And remember, in this world of ruthless efficiency, that ninety-five percent of what a butterfly does is inefficient. (Allegedly.)

Christopher Bryant's tale had a further twist. After his death, the firm of solicitors to which he had enslaved himself sent a bill to his mother, charging her for their time in going to tell her that Christopher was dead. And their justification? The universal petty cruelty; time is money. Shakespeare, ever contemporary, writes in *Troilus and Cressida,* "Time hath, my lord, a wallet at his back." Scholars suggest this could refer to the traditional figure of a lawyer, with a wallet at his back.

Frederick Winslow Taylor, known as Speedy Taylor, was an American industrial engineer who produced a scheme of efficiency at work—Taylorism—which included timing the seconds it took to go to the lavatory. Spending a *penny* was itself costed, for time is money. "In our scheme," he wrote in 1906, "we do not ask for the initiative of our men. All we want of them is to obey the orders that we give them, do what we say and do it quick." He published his *Principles of Scientific Management* in 1911, and his ideas have been enormously influential. In 1915, Taylor's health broke, and he was hospitalized in Philadelphia. In the early hours of his fifty-ninth birthday, a nurse heard him winding his watch and, checking on him at four-thirty that morning, found him dead. Watch in hand.

Recently in the British Library a memo to staff was pinned to the wall: "Please do not start work before the appointed time: 8:36." Today, many work schedules demand employees record what they were doing every fifteen minutes. This minute attention to the minute demeans the mind, fettering its freedom with the shackles of clocks and the watch, the nasty little chain round the wrist.

Electronic time sheets with names like *Carpe Diem* and *Tempo* are installed into computers which employees are required to fill in to show what they were doing at any given minute. One management consultancy divides the day into six-minute units. This can seem ridiculous, with employees spending their time filling in a time sheet saying they were filling in a time sheet.

More seriously, employees are beginning to make protests about "time sheet abuse" and the stress it causes. One mental health nurse has been driven to suicide by it. In this age of "rights" there should surely also be Time Rights, fighting any attempts at the metaphysical enslavery of people's time, arguing for a humane clock, for an integrity of time and respect for the dignity of the individual's hours.

Franklin's lie was peddled in his *Advice to a Young Tradesman* and my, have tradesmen heeded the advice. The "time is money" line is exploited ever more ferociously by businesses, in commerce and global trade, in marketing, advertising and consumerism.

Spatial territories exhausted, businesses expand into time territories, so it is always McMarket day and shops—opening twenty-four hours, open on Sabbaths and festivals—create McTime, a permanent present, obliterating time distinctions, cancelling closed-times, night-times, off-times, odd-times, and nodding-off times. The stealthy language of power is prevalent; there is a potential "conquest" of darkness, businesspeople talk of "colonizing" the night and the "frontier" of dark.

Supermarkets are both a symbol and site of this time. In Britain, under Thatcherism, supermarkets were legally encouraged to see the deep dark vault of night as a potential hole-in-the-wall cash-back opportunity. (Thatcher, appropriately a *grocer's* daughter, was thoroughly in tune with supermarket mean time.)

"Absolute, True and Mathematical Time, of itself and from its own nature, flows equably, without relation to anything external . . . all motions may be accelerated or retarded but the flowing of Absolute Time is not liable to change . . ." This, a sort of prescient motto for supermarket-time, is Isaac Newton's shelf-stacking definition which has blanked the variegated beauty of time ever since he said it. With a superbly appropriate touch, during the Thatcherite eighties, Grantham (proud of its most famous daughter) honored its most mealy-mouthed son by having a supermarket named after him—the Sir Isaac Newton Supermarket.

Companies pollute time, destroying the time-distinctiveness of a weekend rhythm, or exploiting global time zones as does one large American company working an almost continuous day, its data processing done in Northern Ireland, its computer software in Delhi, both cutting wage costs and gaining advantage of being eight hours "ahead."

The globalization of time has the effect of all globalization; the reduction of both the *variety* and the *number* of different times. Once, there were thousands upon thousands of different time zones—as many and as varied as there were communities to choose their own hours, some based on festivals, some on moons and some on money. (Edward T. Hall writes of the Tiv peoples in Nigeria naming the days of the week after things which are on sale in the nearest market, though if you traveled this would vary.) In 1884, it was decided there would be only twenty-four zones. Today, there are largely three time zones, all money-oriented, based on the hours of the Big Three trading blocks, the States, Europe and Japan (and there are huge commercial pressures on other countries to locate themselves within those zones to exploit new power-patterns, of time-advantages and time-disadvantages). Should that same tendency continue, will we end up with just one monolithic time-center, to which all other natural zones would have to bend?

If you drive between London and Heathrow Airport, the closer you are to the airport, the greater the density of corporate clocks; one digital neon says 13:36. Lucozade says it's 13:32. Data General has a clock, as has a mobile phone company, and Polygram's offices say it is 13:48; in advertising their being "up to the minute," these companies target the well-monied Heathrow-bound customers, flattering them with the value of their time by measuring its every vital statistical second.

"Time is Money" permeates advertising strategies and marketing ploys. "Because life is short . . ." runs an advertisement for the communication company AT&T, on the same route, creating imaginary time shortages as merchants have always created market shortages. Watch advertisements play with time; one, for the Swiss fashion watches, showed a man and woman in bed. Captioning him "Always early," and her "Always late," it concluded: "Swatch. Keep in shape." Other recent advertising copy includes: "Keeping the World Up to Speed," "I Want It On Time," "Double Time," "It's a Fast Paced World. Better Keep Up" among others. Not eager to do their work for them, I won't tell you which products they advertise. "Born to Be On Time" is the slogan of one postal delivery company, which suggests that it can actually beat time: "Europe by 8:30 A.M., USA by 8 A.M.," it boasts.

Machu Picchu is an entire temple of sun-worship. One of its major shrines is the Intihuatana, the Quechua word meaning the "hitching post of the sun." It is one of the most revered parts of the site, one of the places where time and spiritual authority are seen to coincide. It refers to a carved rock pillar which tells the "time of the year" so the solstices could be predicted: thus the Inca who was the "son of the sun" was able to confirm his status through association with time itself. All but one of these rock pillars were smashed by the invading Spanish in an attempt to wipe out sun-worship. What the conquistadors left undone, advertisers finished. Attempting to confirm its status through association with time itself, the manufacturer of Cusqueña beer produces an advertisement: *No la pases tan rápido—disfrútala.* (Don't hurry, take time to enjoy it.) Meanwhile, a TV advertisement for the beer was made at Machu Picchu, and one of their cranes smashed into the hitching post of the sun—widely understood symbol of time.

Marketing uses time, with sales tags offering the *latest* dishwashing liquid, selling the interim between the latest and the one just before. Supermarkets use just-in-time practices, whereby products are not held in storage but are produced only on demand, which in practice, in Japan at least, means they are "stored in transit" in delivery trucks which clog up motorways. In the States, the cable giant QVC (Quality, Value, Convenience) has a digital clock on countdown to the end of a commercial in order to create a sense of urgency in viewers to buy while they have the chance. Sometimes time is used simply as an attention-seeking mystery word to capture first your curiosity, then your purse. The Royal Automobile Club in Britain recently put an ad on the underground, with the slogan: "If time is life's most precious commodity, who's controlling it?" with a number to ring. It's irresistible.

Brrring brrring. Brrring brrring. Click.

"OK, then, who *is* controlling time?"

RAC-man: "I've got no idea whatsoever."

(Both laugh.)

"What does the ad mean, then?"

"Oooh, you've got me stumped there."

"I'm writing a book about time, you see . . ."

"Well, time is money. Innit?"

Hhmmm.

If you phone the talking clock, you will find that time is used   **225** as a billboard for a watch company; "At the third stroke, The Time, sponsored by Whojumaflip . . ." In *FutureVision*, the program for the General Assembly of the World Future Society in Washington, the future is a market and companies learn to prepare for its exploitation, it is the sale of time. "Time is power," announces a portentous producer of stationery goods. So "Take control with Day-Timer"—their personal planning system. And, it adds darkly, "if you're not in control of your time, then you're not in control of your life."

The consumer industry scorns anything which is rather *vieux chapeau,* especially, of course, in the rag trade, where time-signals are always used in sales—"fashion, trend, latest and novel." The very words for fashion move in quicker succession than almost any other group of words. "Hip" isn't hip, and "trendy" is not a trendy word for long. As for groovy . . . (Mind you, claims of novelty in fashion are always questionable—Jacobeans were the first punks and Heraclitus was *definitely* a hippy.) In clothes shops, time is reduced to a fashion accessory; "new season stock" invests garments with consumer significance, when the seasonal significance is, for many air-conditioned urbanites, a thing of the past. By calling clothes "this season's clothes," they appeal to a cloaked need for the turning seasons of nature—real changes of the calendar—and reduce that need to its most superficial expression; a change is as good as a vest. There is more: in order to get ahead of each other, American manufacturers decree a new season to have begun earlier and earlier, so spring clothes appear from January, summer clothes in April and winter clothes in July.

Time-is-money underpins the whole philosophy of consumerism: a ubiquitous lie which cheats and saddens the present by promoting a constant preoccupation with tomorrow's purchases today, both saving and spending in the hope of future happiness—a time when you think you will have what you want, but which is always just out of reach, your money never quite catching up with your time. In 1955, the American marketing consultant Victor Lebow argued: "Our enormously productive economy . . . demands that we make consumption a way of life, that we convert the buying and use of goods into rituals, that we seek our spiritual satisfactions in consumption . . . we need things consumed, burned up, worn out, replaced and discarded at an ever-growing rate."

An average American lifetime produces fifty-two tons of garbage per person. More goods and services have been consumed by people alive between 1950 and 1990 than by all previous generations in human history. In consumption society, eating disorders and obesity are rife (Diana, Princess of Consumerism, was princess of both shopping disorders and eating disorders). Overconsumption chimes with overeating, so that the average American buys more possessions than will fit comfortably into the average American house (hence the growth in the storage industry) while the average American eats more than will fit comfortably into the average American stomach (hence the great rise in the outsize clothing industry).

The word "corporation" comes from *corpus*, the Latin for body. A jocular definition of "corporation" is "a protruding stomach." Today, consumers' corpulent stomachs are the personal representation of the political fact of corporations; these gigantic bodies, giant companies demanding overconsumption and planned obsolescence. In planning their future meals, the speed of the last meal's decay is an advantage to the bulimic, in excretion they see new appetites developing. Likewise in planning its future markets, the speed of product decay is an advantage to the corporation, in obsolescence it sees new appetites developing; as for durability, perishable the thought. The very word "purchase" has something stale about it, faintly old, a bit off. It is a word just now on the point of passing its linguistic sell-by date. Why? Because it has an air of durability, a sturdiness that isn't welcome anymore.

Mending things is an act of protest at such corporations, darning socks is a subversive act, a "sew there" in the face of consumerism which demands that you do not mend but replace, alterability represents the alterity of the aptly-so-named *alter*native society and the hippy fashions of patching, of mixty clothes, are in themselves political statements. Organic objects are far more alterable than plastic or synthetic ones and represent life's continuance, changing over time as faces do; old oak or elm tables, for instance, warp and grow; they stretch in a yaw, over the years. New objects, plastic or synthetic, cannot be mended and do not change or warp or grow. They break. Or not. They have a "binary" life, and exist in binary time. Either dead or alive and nothing in between. There is a weird attraction to this binary life for, unlike natural objects which tend toward disorder, modern consumer artifacts tend towards order—the deceitful orderliness of the bin-it factor, the binary is very binnable. Neatly throwing things away demonstrates power over time and tides, a conquest of time by tidiness.

Hong Kong is emblematic of Western modernity's attitude towards time and the world's most famous example of money-making. A place of split-second hurry, a plethora of clocks in public places harry time in every direction you look. People exploit the time-is-money idea ferociously here; every second is made to count: banks, corporations and industry working all hours, day and night, squeezing money out of time. And, lapping Hong Kong on all sides is the great time-giver the ocean. But this is virtually a "dead" sea, foul with *Escherichia coli (E. coli)* and hepatitis, a sink for the sewage and industrial waste tipped in at a speed and in such a quantity the ocean simply cannot survive. Modernity is killing whole seas, making oceans mortal and vulnerable and devastating the vasty deep.

From early capitalism to today's consumerism, the idea that "time is money" is a highly political one—whose time makes whose money and who owns whose time. Usury is the original way to turn one person's time into another person's money. Over five hundred years ago, the Catholic church fought a furious battle with bankers and merchants over loans, saying no one should charge high interest on debts, because time itself is not theirs. As the medieval scholar Thomas Chobham wrote: "The usurer sells nothing to the borrower that belongs to him. He sells only time which belongs to God. He can therefore not make a profit from selling someone else's property."

Today, softly, softly comes the Gombeen Man, the loan shark, the "First" World Bank and the "Third" World Debt. The interest rates paid out far exceed the original loans, so the countries of the South pay for the North's increasing wealth and power. Dirt-poor sub-Saharan Africa scrapes together a billion dollars a month in loan repayments, farming fertile fields to sell food to the overfed West all day, then, at dusk, scratching for the odd dry husk on the midden heap to eat themselves, eating dirt. Every five seconds, a child dies because of this debt. And the churches are, once again, campaigning against the immorality of usury and this debt.

However, the "real foreign debt" is quite the reverse, say the colonized, for between 1503 and 1660, Europe stole thousands of kilos of gold and millions of kilos of silver. This gold and silver could ironically be described as "loans" from the First Peoples of America to the invaders. "Such fabulous capital exports were nothing short of the beginning of a Marshalltezuma Plan to guarantee the reconstruction of a barbarian Europe . . ." says the ingenious Native American chief Guaicaipuro Cuautémoc in an open letter to all European governments, published in "*Revista Renacer Indianista*," asking them to repay the debt. Stating that they will "refrain from charging our European brothers the despicable and bloodthirsty floating rates of twenty or even thirty percent that they charge to Third World countries" they will demand only the devolution of all precious metals advanced plus a modest fixed interest rate accumulated over 300 years. "We inform our discoverers that they only owe us, as a first payment against the debt, a mass of 185,000 kilos of gold and sixteen million kilos of silver both raised to the power of 300." An amount "whose weight fully exceeds that of the planet Earth."

The view which any age has of time has something of the quality of self-portraiture; time is a mirror held up to human nature. Sensitive to that reflexivity, Jorge Luis Borges in *Labyrinths* (1964) wrote: "Time is the substance from which I am made. Time is a river which carries me along, but I am the river; it is a tiger that devours me, but I am the tiger; it is a fire that consumes me, but I am the fire." "*Somos el tiempo, el río indivisible,*" he wrote in "Los Conjurados." (We are time itself, the indivisible river.) Spengler in *The Decline of the West* wrote of time: "We create it as an idea or notion and do not begin till much later to suspect that we ourselves are Time, in as much as we live."

The rigid Order of the sixth-century Benedictines decreed time should be rigidly ordered. Before the twelfth century, before the modern sense of "individualism," when people sat on benches, undivided, likewise time was undivided, the day a long bench of time. Then the period of early individualism, dividing person from person (when chairs began dividing benches), was the period when bells increasingly divided the day into individual hours.

Time, for Chaucer or for the medieval authors of the Books of Hours, was characterized by the deep rural patience which described their own lives, just as today indigenous and sustainable societies see time as sustained. If there was something called the "Renaissance Discovery of Time," as Ricardo Quinones's book is titled, when the forgiving time of Everyman turned to the anxious, guilty time of Faust, what Renaissance men and women discovered more than anything was themselves—the menace which they saw in Saturn was their own. When Walter Raleigh wished to flatter Elizabeth I, he called her "the Lady whom Time hath forgot," for time was seen by the Elizabethans as the devourer, cruel destroyer of youth and beauty, with its scythe, agent of decay and disease: but *theirs* was the cruelty, *theirs* was the "syphilization" which overemphasized youth and beauty while poisoning it. Elizabethans, charnel-minded and skully, had sundials decorated with Death's mocking *memento mori*, which said far more about themselves than about time.

In Shakespeare's *Richard II,* time treats the human as the human treats it:

> *How sour sweet music is*
> *When time is broke and no proportion kept!*
> *So is it in the music of men's lives.*
> *I wasted time, and now doth time waste me;*
> *For now hath time made me his numbering clock*
> *My thoughts are minutes.*

To the seventeenth-century Puritans, time was as drear as themselves; they dressed the calendar in black, forbidding fun and festivals. In the days of empire, when the West imposed its chronopolitics on the world, it argued that *its* time was *the* time, sole, absolute and tyrannical time; it was a portrait of the colonizers, absolutist tyrants with a problem with pluralism.

Now the dominant metaphor for time is money. And, as ever, humanity's relation to time is intricate in its reflexivity. Time is represented as an accountant, mean with its minutes; or global like a multinational; urgent with a greedy haste; time is a *consumer* devouring people; time is running out; time is the loan shark with a knife to your throat; we live at the "end of time" and the "end of history" as all those end-of books so dishonestly portend.

But now as ever, we describe *ourselves* when we think we describe *time*, for it is not time which is the consumer but we who consume everything *including* time. It is we who are "Like a clock worn out with eating time" in the words of John Dryden. Nasty in our haste, the finality is in us; it is we, not time, who are the soulless accountants, the vicious moneylenders making poor nations eat dirt. We are not "eaten up" by time or "burnt out" by time in work, as we pretend, but it is rather we who burn up the earth's time, we are the chronic overeaters, the consumerism is ours, the "greedy institutions" are not performing the offices of time, but are our own inventions. Our image of time is totalitarian, because the totalitarianism is in us, but one writ so large we can hardly read it, that of the rampant and destructive free-marketeering multinationals. The corporation, immortal, invisible, indestructible, unaccountable like a machine made to mouth only *time-is-money, time-is-money*, molds an implicit fascism; mind without morality, money at any price, the market *über alles*.

Time-is-money is a lie. Don't buy it.

nothing regresses like progress

—E. E. CUMMINGS

A postcard from New Zealand.

There is a huge lake here, Lake Waikaremoana, of great significance to Maoris. On one side of the lake there are mossy trees, gnarly-fingered and ancient, and on the other the birds are a hoot of circus sounds. Many tourists take three or four days out, to trek around it. No Maoris live here today.

Except. Should you be lucky, if you stay a night by the lake, you may see a motorboat called *Outlaw* zipping over across the water, bringing Trainer Tate, wild of hair and wide of smile, a Maori freedom fighter whose greeting is to lift his jumper and flash the logo of his black and red T-shirt to show the Maori Independence flag streaming across his wonderfully big belly. The flag also flies over his hut on the lake, his family home for generations. "I've occupied this land, because this is my land and it's where I belong." The government wants him out—he spoils the view for tourists, churning up the lake to catch the odd walker and tell them about Maori cultural self-respect. While some New Zealanders still say the Maoris should "catch up" with whites, that they are "backwards and still behind" on the march to progress, Trainer Tate sees it differently. Apart from the outboard motor, he eschews modern technology. "Your progress is technological progress, and it has not meant progress for us." Just one man, but his message is repeated with the same defiance across the world. My land is not your land. And your progress is not my progress.

A postcard from Colombia.

233

Whistling, riffing, cooing, calling, croaking and trumpeting, the skies are full of birdsong, their fluting song echoes off the clouds. When the birds streel across the skies, they chant the names of the areas they fly over and, by doing so, they create places; or so the U'wa—living in the foothills and cloudforests of the Andes in northeastern Colombia—believe. The U'wa themselves daily sing the world into creation by reciting their myths and place names. Through sustaining nature, they believe, they sustain time itself. (And, who can ever know, that since daytime follows the birds' dawn chorus, maybe birds across the world feel that it is *because of* their dawn shout that day comes.)

At the first invasion, by the Spanish, the U'wa considered their endless and sung hours of time were finished; Berito Kuwar U'wa, an U'wa leader, keeps in his pocket a "clock" insect which whistles on the U'wa hour. "We had lots of hours before the Spanish came." Today, as an earlier chapter noted, the second invasion of U'wa ancestral territory is taking place by companies exploring for oil, the U.S. oil corporation Occidental and (formerly) the Anglo-Dutch Shell. Modern *conquistadors,* they invade so that the West can fuel cars, the powerful symbols and engines of Western progress. "We are seeking an explanation for this 'progress' that goes against life. We are demanding that this kind of progress stop, that oil exploitation in the heart of the earth is halted, that bleeding of the Earth stop," they say. It is now a *cause célèbre.* In an early and definitive article about it, John Vidal quotes Berito saying: "For us, the earth is sacred; it is not for violation, exploitation or negotiation; it is to be cared for, to be conserved." Says a Catholic nun who has spent fifteen years with the U'wa: "I think the traditional U'wa will die if the oil comes. They would enter a great sadness at the end of their land. They will die of sadness."

The end of their land is the end of their time. At the "end" of their land is a sheer cliff, a literal cliff and a metaphoric one, one too sacred to set foot on, too sacred for any footstep, that is, except a final one: for a community of U'wa went to this cliff in the sixteenth century, in retreat from the colonizing Spanish; the tribe put their children into clay pots and cast them from the cliffs, then leapt after them, in a mass ritual suicide. And to this ultimate cliff of fall, frightful, sheer, the U'wa today are proposing to return, for this second invasion, the progress of a second empire, the progress of corporate colonialism. There is now a military presence at the drill-site; a militia trained by the U.S. and Occidental to crush local opposition. (At the time of writing, Occidental announced they were unable to find the oil. This, said the U'wa, was because the U'wa spiritual leaders and medicine people had used traditional rituals to "hide the oil.")

In Peru in the 1980s, prospectors working for Shell spread diseases (such as influenza) to which the local people had no immunity. In one case alone, fifty indigenous people died as a result. The story is repeated again and again across the world. The Surui peoples of Brazilian Amazonia, says Mark Edwards, photographer, must sell their trees as timber to pay for medicine to cure the TB brought in when the highways were driven through: "A process," says Edwards, "which historians are happy to call progress." He has stood where Columbus stood looking at the island of Hispaniola, and where Columbus saw a wondrous landscape of such forests as touched the sky with leaves, Edwards has photographed it as it is today, the desolate stretch of land, a burnt desert, trees cut and forked in charred stumps, "not landscape but manscape," says Edwards. A land has gone to fuel economic progress *somewhere else.*

The Western idea of progress, straight as a Roman road running from the past to the future, is an idea which is so embedded in modernity's psyche that it seems the only possible model of time. It isn't. It's just one idea of time's shape, one out of many, many others, as unusual as it is recent. To the Maori, the past is in front of them so they "walk backwards into the future, facing the past," as Maori writer Witi Ihimaera says. The Quechua people of the Peruvian Andes also "head into the future backwards." The Yanesha of the Selva Central in Peru have the same image. "You face the past because you can see it and keep your back to the future because you can't," says anthropologist Fernando Santos-Granero. For Pawnee Indians in the nineteenth century, life had a rhythm but not a progression. Islam looks backwards to a golden age and an ideal society and the *Qur'an* states that everything that happens has already been told. Arabs look back to look forward, using history both to perceive the future and as a moral justification. Hopi Indians do not recognize the Western divisions of past, present and future, but instead divide time into "that which is manifest" (more objective), and "that which is beginning to manifest" (more subjective). Real things, being manifest, lean to the past while, say, feelings or hopes are things becoming manifest and so lean to the future. (We live forwards, but we understand backwards, in Kierkegaard's formulation.) For the Arakmbut, time is conceptualized as "movement which passes human beings by, leaving them to grow older and change. The metaphor is thus not of people drifting in the river of time, but watching from the banks as it goes on its way. The passage of time is measured by how things change," writes Andrew Gray in *The Last Shaman*. Among the Inuit of Baffin Island, the term *Uvatiarru* means both "long ago in the past" and "far ahead in the future." To the Mayan civilization, cyclicity was all and everything which had ever happened re-enacted itself in a circle of time such that every two hundred and sixty years, history was thought to repeat itself.

To many cultures, time is as cyclical as the fundamental spirit of nature. Hehaka Sapa (Black Elk) of the Oglala Sioux tells how everything in his worldview is done in a circle:

> The Wind, in its greatest power, whirls. Birds make their nests in
> circles, for theirs is the same religion as ours. The sun comes

forth and goes down again in a circle. The moon does the same, and both are round . . . Even the seasons form a great circle in their changing, and always come back again to where they were. The life of a man is a circle from childhood to childhood and so it is in everything where power moves.

An old Lakota shaman circles his sacred pipe to the four winds, explaining that the circle is the symbol of time, because the sun, moon and seasons all circle the world. The year is defined as the circle around the edge of the world. In stark contrast, Western modernity's view of time as the line of progress is, in theory and in practice, opposed to the cycles of nature, not aligned with them.

Although many shapes of time are geometric—most obviously the circle or line—they are not all so; Islam describes "packets" of time and the Navajo (who traditionally do not believe in the future because it is not in existence) have "pulses" of time. There is a cyclical dimension here; Navajo "sun-time" is based on the repetition of recurring events which could make all times interchangeable in ritual.

Progress is only an idea, a mental construct, but is treated as if it had the status of concrete fact, as if the march of progress had a sort of absolute inevitability and preordained certainty. There is a "cultural genealogy" in the West's view of itself, particularly in the U.S., that ancient Greece begat Rome, Rome begat Christian Europe, Christian Europe the Renaissance, the Renaissance the Enlightenment, the Enlightenment political democracy and the Industrial Revolution. Then "Industry, crossed with democracy, in turn yielded the United States, embodying the rights to life, liberty, and the pursuit of happiness," wryly comments Eric Wolf in *Europe and the People Without History.* "Such a developmental scheme is misleading . . . because it turns history into a moral success story . . . a story of how the winners prove that they are virtuous and good by winning," equating goodness with success as W.H. Auden said. In the immediate aftermath of the destruction of the World Trade Center, the *New York Times* prints the liturgy: "the real clash today is actually not between civilizations, but within them—between those Muslims, Christians, Hindus, Buddhists and Jews with a modern and progressive outlook and those with a medieval one."

The dogma of progress was aptly expressed in the concept of the "Manifest Destiny" of one race to subdue or destroy others and to do so with the moral backing of an off-ground god, the Manifest Destiny of the whites to annihilate Native Americans on the ground. "Progress is God," cried William Gilpin, governor of Colorado, and the "occupation of wild territory . . . proceeds with all the solemnity of a providential ordinance." Like progress, Manifest Destiny is invoked to justify one people expanding their territories no matter who stands in their way, progressing across space and into the future.

Progress is a specific idea; Western, money-oriented, technologically biased and racist in its history and its effects, but it pretends to a universality, so that all peoples must be made to define and embrace progress in exactly the same way. The U'wa don't. They know their progress is not served by oil companies destroying their land for Western progress. Your progress is not ours, as Trainer Tate says.

Progress pretends to be an absolute good because it is defined by those it serves well; the rich, the politically powerful, all types of colonialists and ideologues. Ask those whom it serves badly and they will tell you that the engines of "progress" have justified the destruction of lands and peoples of the land, from racism to land thefts, from pollution to genocide.

Progress is a one-word ideology and one which has suited both the Marxist worldview and also that of the Neoliberal far right of multinationals and global free-marketeers. Its dubious ideology is, however, far less explored by these rightist forces of capitalism, who push it as an inevitable, neutral "good," silencing its highly political character. Devotees of progress can be sneeringly contemptuous of anyone who would dare speak against it. Those who stand in the way of progress are called ridiculous, backwards and reactionary. They can be tortured for it. Sure-yani Poroso was tortured by authorities in Bolivia for campaigning for his tribe, against neoliberalism, land thefts, water privatization and the TransAmazonica highway. "For doing this I was called anti-progress. When I was being tortured, they said, 'You are a shit. This is what you get for fighting progress.'" When the word on a torturer's lips is "progress" you see the vicious side of its ideology.

This chapter offers some reasons for making a stand against this so-called progress, to see it for the destructive ideology it is and to ask whom it serves.

Progress is pictured as a march on the road to the future from the past, as a walking—or driving—away from somewhere. This is a crucial image, for the abstract idea of progress depends, and has always depended, on first, the rejection and then, the destruction of place. Missionaries commenting on the Toyeri people in Peru said: "It is important to take the Toyeri out of their jungle environment . . . in order to introduce them, gradually, on the road to civilization." The key inventions of technological progress are thought, by one historian at least, to be the wheel, the stirrup, Portuguese ship-design and the combustion engine, all engines for shrinking distance, for speeding away from place. (In the 1850s, a Swedish visitor to the U.S. Patent Office noted that most of the machines registered, nearly fifteen thousand of them, were inventions "for the acceleration of speed and for the saving of time and labor.") As the word "invention" has overtones of travel, meaning literally "a coming-upon-something," so the very word "progress" has nomadic color, implying a journey. This is, perhaps, part of its unconscious appeal, for even in our virtually sedentary society, the nomadic instinct, both friend and foe, is with us still, restlessness in his feet, horizons in his eyes. No longer nomadic, there is an urge to move through territories of time, when the human can no longer move through space.

The notion of progress has shadily been around for hundreds of years, but it only really gripped the Western mind in the nineteenth century. Manifest Destiny began to be bandied about from the 1840s and, over in England, listen to Macaulay in 1835: "We are on the side of progress . . . The history of England is emphatically the history of progress." And, in those years when the march of progress picked up speed with steam engines and combustion engines, Macaulay pumps out the sound track:

It has accelerated motion; it has annihilated distance; it has facilitated . . . all despatch of business; it has enabled man to descend to the depths of the sea, to soar into the air, to penetrate securely into the noxious recesses of the earth, to traverse the land in cars which whirl along without horses, and the ocean in ships which run ten knots an hour against the wind . . . it is a philosophy which never rests, which has never attained, which is never perfect. Its law is progress.

And *The Origin of Species* was just around the corner. Published in 1859, it ends famously: "As natural selection works solely by and for the good of each being, all corporeal and mental endowments will tend to progress towards perfection."

Ambivalent about progress as an innate good, now embracing, now spurning it, Darwin disliked social applications of his theory, yet his ideas were widely used to maintain that progress *was* improvement, to equate later with better, to see succession as success. And it had two particularly nasty applications, first seeing

Homo sapiens as superior to all other species and, second, seeing the white race as superior to all other races.

Humanity's belief in its superiority has a long history; the ancient Greeks believed the universe had a built-in design of progress in which everything non-human existed for the benefit of humans. Aristotle considered all animals were made for the sake of mankind. From *Genesis,* the Judaeo-Christian tradition taught that mankind should subdue the natural world. Thomas Aquinas thought that plants were made for animals and animals for people. Freud thought that human progress could only be won by conquering nature, "going over to the attack against nature and subjecting her to human will." Anthropologist and occasional humorist Marshall Sahlins comments: "We are the only society on earth that thinks of itself as having risen from savagery, identified with a ruthless nature. Everyone else believes they are descended from gods. . . . Judging from social behavior, this contrast may well be a fair statement of the differences between ourselves and the rest of the world."

Theories of progress seem always to have appealed to societies seeking dominance, over nature and over other peoples—graphically illustrated by the way progress was seized on by European imperialists. Darwin's evolutionary theory gave them both a murderous motive and a coldly clean conscience. In the *Descent of Man* (1871), he writes, "The civilised races . . . will almost certainly exterminate . . . the savage races." Genocide in modern history did not begin with the Nazis but with the colonialists and the ideology of progress.

In his recent book *Exterminate All the Brutes,* Sven Lindqvist writes with savage anger of "progress" resulting in racism and murder in Africa; it was "a progress that presupposes genocide." Giving a lecture in 1838, "On the Extinction of Human Races" the anthropologist J.C. Prichard portrayed the "inevitability" of progress. "The savage races," he said, could not be saved. The aim must be simply to collect as much information on their characteristics in the interests of science. A pattern of thought still followed today; the Human Genome Project, for instance, which ignores the extinction of indigenous tribes in order to concentrate on taking genetic samples from them—and all in the name of scientific progress.

"After Darwin," writes Lindqvist, "it became accepted to shrug your shoulders at genocide. If you were upset, you were just showing your lack of education." And extermination was not just a result, it was a pleasure. Even more delightful than pleasure, to the Victorians, it was a duty. Lindqvist reports how German philosopher Eduard von Hartmann spoke of "the death struggles of savages who are on the verge of extinction," and von Hartmann commented that "the true philanthropist, if he has comprehended the natural law of anthropological evolution, cannot avoid desiring an acceleration of the last convulsion, and labor for that end." W. Winwood Reade, in his book *Savage Africa* (1864), writes of the probable extinction of Africans: "We must learn to look at this result with composure. It illustrates the beneficent law of nature, that the weak must be devoured by the strong."

In the 1890s, in evidence of the murder of people in what is now the Democratic Republic of Congo, baskets of hands were given to the commissioner as signs of the victory of Western progress.

Progress. In the name of progress, murder was philanthropy. "We know," says Lindqvist. "It is not knowledge that is lacking." We have always known "what outrages have been committed and are being committed in the name of Progress . . . It is not knowledge we lack. What is missing is the courage to understand what we know and draw conclusions."

.   .   .

Technological progress has made this the age of the Amplified Man, moving ever faster away from place. Herbert Girardet, the filmmaker and writer who coined the phrase, pictures the Amplified Us in his book *Earthrise.*

> Instead of muscles, we deploy motors; instead of legs we have acquired wheels; our hands are the shovels of bulldozers . . . Thus we have become both man and machine . . . Every year, each of us swallows many tons of coal and oil. Our breath is no longer just what we exhale from our mouths, but also the exhaust fumes spewed out by our cars, factories and households. We excrete the wastes of our own bodies as well as the poisonous discharges of the machines with which we have become fused.

Progress proceeds by annihilating distance and the politics of transport express the politics of "progress." Take the car, worldwide symbol of today's "progress." But whose? The U'wa will die so that oil companies can bring petrol to the West. In Nigeria, Ken Saro-Wiwa and other Ogonis were killed for it. Even within one country, the car's progress is political, increasing inequality; the swift progress of cars forces delays on cyclists and pedestrians. Most transport systems serve the progress of the wealthy, at the cost of the poor, the very young or very old, who suffer bad public transport, increased accidents and increased pollution. Progress is two-faced; it has a lovely smile for the powerful and a cruel sneer for the poor and underprivileged. Such "progress" is also sexist; men have far more access to cars—clean, speedy and safe— while women spend hundreds of thousands of hours escorting children in dirty, dangerous environments; waiting alone in frightening, dark bus stops or train stations; or walking along pavements being solicited by men progressing in cars.

A postcard from the PanAmerican Highway, which, to aid the progress of cars, will carve up the land of the Colombian Embera tribe. To road builders, latter-day colonialists, the Highway means progress. To the Embera, whose forests are threatened by it, the road means cultural genocide bringing, as it does, disease and destruction.

If you want to know something about the march of progress, ask a man who walks. Satish Kumar, editor of the international magazine *Resurgence,* was in his Indian boyhood a Jain monk. One of the Jain principles was not to use any technology he couldn't make with his own hands. No soaring into the air, no penetrating the earth, no traversing the land in cars. An Unamplified Monk. As a young man, he began a walk. He walked from Gandhi's grave in Delhi to Moscow, Paris, London and finally Washington.

His journey was undertaken, he says, for world peace, in the widest sense of the word; peace between nations and "making peace with nature which technological progress has made war with." If technological progress destroys place, Kumar's sense of progress—"a pilgrim's progress"—*enhances* a sense of place.

> In every place, I was *there,* I wasn't making progress to *get* somewhere but to *be* somewhere. It's about relishing the place where you are. Technological progress is always you-are-not-where-you-are. That's why walking is so important—place is sensuous, so you feel the air and you hear things. The speed of technological progress destroys sensuality.

In contrast to the sensuousness he describes, there is a seedy promiscuity to car progress, its mobility achieved without effort or emotion, without feeling or relationship to the place it is in, without commitment, responsibility or love.

If progress has always meant a rejection of place and of land, it's hardly surprising that some of its cruelest effects have been on nature—from pollution to global warming and treelessness. While organic waste-processes create time itself, furthering the reproductions of time as manure aids the rejovialization of upthrusty new shoots, non-organic pollution, by contrast, produced by the engines of progress, snips regenerative time in the testicles—synthetic waste, which cannot be reabsorbed, engenders not time but toxicity.

Global warming—much of it a result of mankind's urgent "progress" in transportation—has altered nature's time to the extent that in Britain spring in the late 1990s sprung a week earlier than it did in the 1970s. (In December 2000, in complete defiance of the date, frogs spawned, butterflies fluttered by and swallows were singing.) Our progress may provoke bluebell-progress, for bluebells, in their violet, shy asking, may have to "move" out of England to find congenial weather for their needs and may only flower in Scotland. (Maybe only Scotland deserves them.)

Trees may be badly affected by this transitional and inconstant weather, for they, after all, can't move to a more comfortable spot. Certain types of trees along the Canadian-American border, for instance, though they have slowly migrated for thousands of years to follow gradually shifting temperature zones, will be outrun by global warming. A rise of one or two degrees over thirty years would require them to "move" at five kilometers a year. Trees can't run. Ah, but they can worry. German foresters have coined a term "anxiety shoots" for trees damaged by air pollution and compensating by producing fresh shoots jutting out of their trunks.

In Hindu thought, trees were the ancient philosophers—the cosmos, in Hinduism, is a great tree. Tree-respect has been almost a universal religion, in all ages, so to Buddhists, the *pipal* tree is a symbol of wisdom. In Chinese, Egyptian, Japanese, Scandinavian and Teutonic lore, there is reverence for a tree of life. In Islam, the tree of blessing represents illumination and wisdom. It is as if there is an innate knowledge, rooted in the body, that a tree is a friend to humanity. Western science has shown they are the oxygenating lungs of the Earth, vital to the ecosystem and home to millions of species. Trees are indeed trees of life. Time is different in a treescape; some trees are thought to have stood for hundreds— even thousands—of years. The Bristlecone Pine "Methuselah tree" in California's White Mountains is over 4,700 years old, a tree-history quite humbling for the human of today. And trees don't just "last" passively over time, they "create" time by creating breathable air.

In protest against commercial logging in the Himalayas, women in the 1970s formed the Chipko movement, hugging trees to prevent them being cut down. (*Chipko* means "to hug" in an Indian language.) For trees stand in the way of progress, as the Chipko women, the Penan, and the British road protesters, who build and live in tree houses to protect the trees, know to their cost. So fire up the chainsaws and burn down the forests, fell the trees for roads and put a match to it all, cut it, log it, sell it for cattle-pasture for burger culture. Living trees stand in the way of progress, while the felling of trees creates economic progress. And don't ask why poisonous fog suffocates six countries of Southeast Asia; and don't ask why fires in the forests of Indonesia may burn for ten years; and don't ask why, with the loss of the rainforests, or cloudforests, all the things which used to live in these forests are threatened—the winking, teeming, burping, mischief-making, love-making, tree-climbing, armpit-scratching, crawling, laughing, slimy, furry, juicy, bug-eyed, big-eyed, one-eyed, hundred-legged, leaf-look-alike life—Darwin's "endless forms most beautiful and most wonderful" are rapidly being driven to extinction.

No one knows how many species there are, nor their rate of extinction, but the estimated figures are a speedometer for mankind's journey of technological progress. In the whole of the eighteenth century, when Homo sapiens And Son progressed by footpower and horsepower, kicking up just a little dust on the path, it is thought some twenty species became extinct. In the nineteenth century, with people now driving a steam engine of progress, an estimated eighty-two became extinct. Early in the twentieth century, with the technological progress of the combustion engine picking up pace, one species a year became extinct. In 1990, age of rocket-power, one species became extinct every five hours and by the end of the century one species was thought to become extinct every twenty minutes.

Talk to environmental scientists (such as Dr. Matthew Griffiths, professor of environmental physics at the University of Connecticut) about "progress" and they go chalky pale and plead to be allowed to explain the idea of the exponential and its applications. Take two queens. Once upon a time, these two queens spent their afternoons playing chess, the loser paying a forfeit to the winner. One day, the winning queen asked for her prize thus: the loser was to give her one grain of rice for the first square on the chessboard, two for the second, four for the third, eight for the next, doubling the grains each time. The losing queen smirked up her sleeve at this trifle and began. But the slamming spirit of the Exponential (embodied in the winning queen) had pulled a fast one. Keep doubling the grains of rice and by chess-square number sixty-four you will have bankrupted a kingdom—and a world. Economic activity *plus* pollution *plus* population growth *plus* the use of the earth's resources are increasing exponentially. "It is as if we are driving faster and faster at the wall," says Griffiths. E.F. Schumacher wrote, "We must study the economics of permanence." Who disagrees? Economists, mainly, with a (highly political) belief in encouraging unlimited growth and a misty faith in doing without nature altogether.

December 1987. A postcard from Stockholm. And the Nobel Prize for Economics goes to . . . pause . . . Robert Solow of MIT for his theory of economic growth based on the dispensability of nature. Progress-thinking, ever on the move away from place, is here shaking the last dust of natural earth off its engineered boots. As Solow said in real ignorance: "The world can, in effect, get along without natural resources, so exhaustion is just an event, not a catastrophe." And he proved it with a calculator. An answer to Solow came shortly afterwards, with the ill-fated Biosphere II venture, the 1991 attempt to build a self-sustaining, artificial, enclosed ecosystem in the American Southwest, costing two hundred million dollars. There were massive problems almost from the start. The oxygen level dropped so much they had to open the Biosphere up. Paul Ehrlich of Stanford University says: "Crazy ants went crazy: nineteen of twenty-five species of vertebrates went extinct. The people could barely survive . . . because all sorts of things we take for granted are supplied to us, easily, by nature, but we don't know how to replace them."

Solow was basing his Neoliberal far right economic beliefs on the ideology of "progress." Compare his thoughts with the writers who, under Stalin, were commissioned to write in honor of technological progress as a defeat of nature. One, Vladimir Sasubrin, wrote:

> May the brittle, green breast of Siberia be clad with the concrete armour of cities, fortified by the stony mouths of factory chimneys, and shackled by railway tracks! May the *taiga* be turned into ash, may it be logged and the tundra be trampled. So be it, because it is inevitable. Only on foundations of concrete and iron girders can the comradeship of all humanity, the brotherhood of man be erected!

The price of such ideology is, as ever, paid by the underprivileged; today, in Eastern Europe, there are terrible environmental problems of damaged soil, poisoned rivers and polluted air. In the industrial centers of Poland and the former Czechoslovakia and East Germany, pollution can be measured in human time: people die, on average, five years younger than in nonindustrial areas.

Progress has an enormous appeal for ideologues of both left and right and for ideologues-of-technology, all of whom use its highly political character, while pretending that it is non-political. "So be it, because it is inevitable," said Sasubrin. So be it because it is inevitable, says Neoliberalism in the face of the U'wa tribe's mass suicide. And as cases of mysterious sickness and contamination rise in a sad, bullied, underprivileged Welsh village where Monsanto sited a huge factory, the unconcerned authorities seem to shrug and say the same. So be it because it is inevitable. There's something else. Progress, described as inevitable, is thus not only treated as non-political but also subtly denoted as "only natural," as if it works like a law of nature, which is perverse indeed, considering how modernity's progress destroys nature.

Leading scholar on modern Tibet, Tsering Shakya, described how the Chinese invading his land were driven by the "Marxist ideology which emphasises the idea of material progress." The whole Chinese system in Tibet "is perpetrated in the name of progress . . . to be forced upon 'backward' people who stubbornly refused to surrender to the march of progress. Such people, alas, are the inevitable victims of modernity." A Tibetan monk, Palden Gyatso, in his autobiography, remembers a Tibet without machines. "The Land of Snows had no need for the wheel." By contrast—and fittingly—the first thing the Chinese did on occupation was to build roads for the wheels, the tanks and the engines of progress. The Tibetans had the prayer wheel of tradition.

Time, Forward! was the name of a Five-Year-Plan novel by Katayev; time whipped by ideological urgency and dominating technology. "If a mountain is in the way, we'll move it; if a river flows the wrong way, we'll divert it." Trotsky, here, sounding uncannily like the unreformed World Bank, the IMF or other totalitarian capitalists. It seems appropriate that at the Narmada dam and all over the world, the finest critics of progress are people of the land. For, as in the first colonizing era, it was the non-white races who suffered from the European definition of Progress-As-Genocide: so in this era of corporate—or Marxist—colonialism, it is people of the land who suffer from modernity's progress away from nature.

It's a clash of two cultures and two mind-sets; one characterized by off-ground ideologies (including both rampant capitalism and far-left views) and by the idea of the Temporary—the flashy mobility built in to Western cultures. The other is based on permanence and a love of place. "All the knowledge of our people is based on a permanent relationship with the places in which we live. The Indian territories are not only physically but also culturally located," says the passionately eloquent Ailton Krenak, president of UNI, the Association of Indian Nations based in São Paulo:

> The blood of the forest is the sap, it runs through the tree until it meets the leaves. Together with the wind, it makes a song. A son of the forest hears these things and memorizes them. In the village he will sing for his people and if they approve of his song it will be incorporated into what you could call the musical history of those people. You could arrive in that place ten thousand years later and after five or ten days you would hear the relatives singing the same song.

Ten thousand years. Till he might have said with Tennyson, "Let us hush this cry of 'Forward' till ten thousand years have gone." (In the West, where the music industry is so transient that "Yesterday" was written in a completely different age from ours, a ten-thousand-year-old song is all but inconceivable.)

A postcard from Norway. It is 1981, and the Norwegian government has just deployed its largest contingent of military force since the Second World War. Against who? The Sami—the Lapps, as they do not like to be called—who were protesting against a hydroelectric scheme whose effects on their reindeer herds would be incalculable. This is Progress, thundered the government, most unnorwegianly; this is *development*.

A postcard from the Karen, in Northern Thailand, where Jim, a British man, teaches the women to knit baggy, ugly Western jumpers. Some of them are bobbly and some have sequins on. They do it very, very badly. I almost offer to help except I would—just—have done it worse. (The Karen, needless to say, have for centuries made beautiful woven textiles.) Jim teaches them "how to use their time" by which he means knitting for eight hours a day for a cash payment. He likes the women to come to him on their knees for their wages. "It's traditional," he says. Before this little Lord Jim came, the Karen hardly used money and he is pleased with his impact, gesturing to explain, extending his hand downwards to an imaginary chain of Karen people on a lower rung. "I want to lift them up from their position up on to our own level." Progress means cash, to him, and he wants them to be on the Western path of economic progress, knitting all the way to the bank.

"The moon rises differently and the sun sets differently now," say the Weyewa people of Sumba, Indonesia, explaining how economic progress has altered the very character of their time.

A postcard from Sarawak, where Datuk Amar James Wong Kim Min, who is a "well-known and effusive businessman and politician," and President of the Sarawak National Party, according to the back of his book of poems *Buy a Little Time,* a book which is, the blurb on the cover tells you, "vibrating with humanity and resonating with the highly strung cords [*sic*] of sentimentality through every page." He writes a poem to the indigenous Penan, urging them to "cross the Rubicon!" towards progress— "join our civilisation," "cast aside primitive Tradition." He would, like Lord Jim, offer to "lift a helping hand" to raise them to his higher level of development. He is a keen golfer. Let the Penan give up "subsisting on Blowpipes" and join him on the golf course of development.

Little Lord Jim and little Jim Wong want to see the Karen and the Penan "developing." This is a key component of contemporary theories of progress which retain all the racism implicit in "evolutionary development," all the fake assumptions of the abstract inevitability of progress and all the Western crassness of only measuring wealth in cash. "Developed" implies the primacy of a first world as opposed to a third, and "advanced" economies as opposed to "stagnant" or "subsistence" or non-cash economies. It implies that countries of the North are further ahead and therefore superior to countries of the South and that there exists in some preordained way a *road* of progress to be either further advanced or backwards and behind on. A *race* of progress to be first in, so indigenous peoples must catch up. The neoliberal analysis of Mexico is that there are two nations, NAFTA-sponsored Mexico and "*el otro Mexico*," the other Mexico, backwards and left behind. (The proferred solution is increased debt.) In Indonesia, indigenous peoples are officially characterized as "people who are isolated and have a limited capacity to communicate with other more *advanced* groups, resulting in their having *backward* attitudes." In 1998, a history book used in Peruvian schools describes indigenous people as *salvajes* who are *atrasados*, savages who are backward.

When President Truman put forward the idea of the "underdeveloped" world in 1949—a world which must be industrialized, technologized, capitalized, consumerized and indebted to Western banks and Western experts—what he didn't know was how underdeveloped other peoples thought Americans and Europeans were, how disastrously immature the West was thought to be, in subtlety of communication, in social living, in earth-lore; how undeveloped in sensitivity, in sensibility, in reciprocity, in gift-giving, in kindness and in pity.

The Kogi Indians live in virtual seclusion in the Sierra Nevada mountains of Colombia. In 1988–9, the Kogi *Mamos,* their priests and judges, took the unprecedented step of coming out of isolation to warn of the damage of Western progress, saying that the climate was being changed, that crops were suffering drought. They prophesied that, by removing trees and plundering minerals, the world would heat up and that if Westerners, whom they referred to as the "younger brothers," did not change their behavior, the world would die.

This seems to be mysteriously widespread; both the warnings of progress which indigenous peoples are voicing and the idea of Westerners as "younger brothers." Berito Kuwar U'wa spoke of the U'wa being "older brothers" responsible for looking after the Earth. Both the Hopi and the Arhuaco Indians use that same image. Across the world, according to Karen myth, the "white man" is the "younger brother" of all the races. And this younger brother is given uncannily similar characteristics; he is brilliant but dangerously immature, intelligent but arrogant, adaptable but cruel, able but greedy, the inventive manufacturer of boots he is too big for, the wizard mathematician of halves which he is too clever by, the crafty carpenter of bargepoles which he shouldn't be touched with.

But portraiture, like history, is commissioned by the victors and the West has chosen the more flattering painting; not of younger brother, but of advanced culture, for in the chronopolitics of the twentieth century, advanced nations zoom ahead, modernized, developed and progressive, while the rest of the world trundles along with a wooden-wheeled wheelbarrow, backward, undeveloped and stagnant. The idea of the Old World and the New World is a lie too: the Younger Brothers are of the New World, and the Old World properly belongs to the indigenous peoples.

Cardinal John Henry Newman once said: "Progress is a slang

term." For many, it is a four-letter word. Vandana Shiva spits on the shoe of progress and development—including, not surprisingly, the Narmada dam.

> What is currently called development is essentially maldevelopment, based on the. . . . domination of man over nature and women. The economic "growth" that the masculinist model of progress has sold has been the growth of money and capital based on the destruction of other kinds of wealth such as the wealth produced by nature and women.

She argues instead that "steadiness and stability are not stagnation, and balance with nature's essential ecological processes is not technological backwardness but technological sophistication. . . . Chasing the mirage of unending growth, by spreading resource-destructive technologies, becomes a major source of genocide, the killing of people by the murder of nature is . . . today the biggest threat to justice and peace."

The association of injustice with progress has a long, long history. In 1879, Henry George, the American writer on political economy, wrote *Progress and Poverty* which is as relevant today as then. Where progress is most evident, he writes, "we find the deepest poverty, the sharpest struggle for existence, and the most enforced idleness . . . This association of poverty with progress is the great enigma of our times." Today, look no further than the way the development loans to so-called Third World countries, for their so-called progress, have merely increased the wealth of the rich by impoverishing the poor. George: "While modern progress . . . makes sharper the contrast between the House of Have and the House of Want, progress is not real and cannot be permanent."

·   ·   ·

Today, the act of standing is virtually a political statement, a political *stance*. Trainer Tate "stands his ground" both in his message and in his occupation of his land. The Chipko women stand hugging their trees. Road protesters in Britain stand in front of the bulldozers. The Haida people of Canada in the 1980s blockaded logging operations by blocking a road. And Ken Saro-Wiwa stood up to Shell. Progress is emblematically represented by roads, Macaulay's engines, the tanks of the Chinese in Tibet, the Western cars and oil industry which will kill whole tribes: these are metaphors of modernity's journey, ever out to destroy land and peoples of the land. The U'wa stand at the edge of their cliff so that car drivers can fill their tanks for an ideology of progress quite as vicious, though by no means as honest, as the Chinese tanks in Tibet.

Ken Saro-Wiwa, leader of the Nigerian Ogoni before he was murdered, said of the Ogoni land: "The land is a god and is worshipped as such." Too bad. Progress means the engine is a god and worshipped as such, and land must be sacrificed to it. Speed is a god and must be worshipped as such. Beneath the Ogoni lands were vast oil finds; Shell and Chevron took a hundred billion dollars' worth of oil to fuel *their* economic progress, not the Ogoni's. And left Ogoniland destroyed. The oil spills polluted the rivers from which the Ogoni people drank and the mangrove swamps in which they fished. "Where are the crabs, periwinkles, mudskippers, cockles, shrimps?" asked Saro-Wiwa. Nowhere. The land, he said, was made infertile by acid rain. Oil covers roofs, trees and grass. And Saro-Wiwa described how what once was the sound of water in the streams had turned to a silence of oil sticking noiselessly to rock. Shell has advertised itself "Moving at the Speed of Life." For Saro-Wiwa, Shell spelled death and he was killed for his protest against them. He was depicted, in Steve Bell's famous cartoon, hanged by a Shell petrol hose, tied with the noose of a corrupt government until a new silence—the silence of oil sticking noiselessly to death—fell.

Enter Daniel Zapata.

Zapata, spokesperson for the Hopi, arrives for supper, all tattoos and headbands flying, crooning with his walkman and flamboyant with righteous anger. We talk "progress." In Hopi myth, Zapata says, the duty of Native peoples is to take care, while the duty of the white peoples is to be creative and innovative, but the "little lost white brother, the youngest of the bunch" has used his power to destroy the earth with his technological progress. He links this directly to the success of corporations in their march of political progress. "This is an age of *corporate* government and *corporate* democracy. The idea of a people's democracy is disintegrating."

Talking of nature and of First Peoples, his hands move in figures of eight, the figure of infinity. Sustainability has him drawing circles in the condensation on the window. By contrast, talking of corporate greed for economic progress which results in environmental catastrophe, his fingers meet like a stiff spire, oppositional and finite. "The natural law of progress is like infinity. It's the complete opposite of a mechanical law of progress or the corporation law of progress."

A postcard from Brazil. On the Brazilian flag are the words "Order and Progress." For the Wayapi people of the Brazilian Amazon, according to anthropologist Alan Tormaid Campbell, progress means that their lands are taken, their forests "logged and burnt for cattle-grazing, polluted by schemes for mineral extraction, flooded by the building of monstrous dams." The Wayapi people are, in his beautiful phrase, "poised on their last seasons of integrity."

Amilton Lopez ("that's the name they gave me in church, but my indigenous name is Ava Pykavera") is a Kaiowa chief, of the Guaraní-Kaiowa peoples of Mato Grosso do Sul, in Brazil. He is in London to talk about progress, for land that used to belong to his tribe is now being taken by cattle ranchers for the progress of burger culture—and his tribe is devastated. As we talk, he suddenly reaches out to me and grips my arm. It tingles. He holds my wrist. "If I had earth on my fingers," he says, rubbing his fingers up and down my forearm, "if I rub it on your arm, your white arm, you could see it. But if I rub it on my arm, you couldn't see it." He is *of* the earth, he says, there is an absolute inseparability of the land and the people of the land. He looks around the room we are in—all gray plastic and computers—"All this, files and paper and carpet doesn't make any sense to us." But as their land is taken away, suicides are increasing, particularly among young girls. There were forty-three suicides in 1995 alone. "People are killing themselves because they have literally no more space to live in. Everything is finished. The forest is finished so our culture is finished because our culture is in the forest, in nature, in the environment."

In one Kaiowa area, people have threatened, like the U'wa, to commit collective suicide if their land is taken from them. "We *are* the land," he says, willing me to understand that profound simplicity, and he can't say any more. It would have broken your heart.

None of them can live far enough away from Western progress for, in a conceptual sense, there is no place left on earth which Western progress does not shadow, this progress which is such an enemy of place, of land and enemy of the people of the land. Only by dying can they move far enough away. This is the last part and the last part is sad.

A clearing, a desert plain. Against the skyline, if you knew the land as they did, you could see, in your mind, the black silhouette of where the trees used to be. One guava tree remains in a deforested waste and from a branch a twelve-year-old girl in a sundress hangs dead in the moonlight, dying in a vanishing time.

10 · A TEFLON TOMORROW

You've no idea how pleasant it is not to have any future . . . It's
like having a totally efficient contraceptive.

—ANTHONY BURGESS, *Honey for the Bears*

If time can be divided into past, present and future, each of these
has a location. The past is underground; it is the place of burial,
of geological history, of archeology. The present is at ground-
level; those who walk the earth today. And the future is off-
ground, its site is the sky, its castles are in the air. In terms of the
human mind, if the past is the place of memory and the present
is the place of perception, the future is the place of imagination,
of thought, of dreams. The practice of using dreams to interpret
the future is common, anthropologists such as Philippe Descola
say, to many pre-modern societies. He writes of the Achuar of
Ecuador that no one goes out hunting without having had a
dream presaging it. Hugh Brody, in *The Other Side of Eden*, describes Athapaskan societies of Alaska using dreams to foretell—
or perhaps more accurately foreguide—a hunt.

Progress appeals to this future orientation, for progress—intrinsically an enemy of place—ever dislocates itself from the
muddy, earthy today towards a tomorrow in the sky and in the
imagination. That is why progress is still so appealing, even when
the negative effects are everywhere, from Ogoniland to the hole
in the ozone layer to genocide, because the habitation of progress
is in the human mind, wherever mind is—the appeal of progress
is that of imagination itself. Its fascination is in the metaphysical
world as opposed to the physical, not the factual but the meta-
fantastical, the meta-place of the mind; insatiate for fire not earth,

the burning ambition of the mind to superbia where matter doesn't matter, because here is the vertigo of sheer mind that thinks best alone; the disconnection of genius, the dislocation of the brilliant—the brilliance, though, of both genius and madness.

Progress has the irresistible, heady draw of fire. You can deny the positive effects of progress. You can deny the good of technology. You can deny all its results but you cannot deny its appeal. Flame. Incandescence. The progress of technology, of virtual reality or space research, of science-whatever-the-price, has a quality of flame in flight, the thought pursued though we die for it, death-or-glory, challenging the stars and stabbing the heavens with the arrow of progress shot by sheer visionary thinking, flaming its line of fire across the mind's sky.

But its appeal is deceptive. Its fantastic appearance is not its reality; for the sake of such incandescence, Indonesia and Brazil are actually on fire. All over the world, forests burn with fires which can be seen from the moon—while the moon, once symbol of pure imagination, is now symbol of political and commercial ambition, and reached, of course, by fire. Shell flares Ogoniland until there is no more night, and in cyberspace, people "flame" other users with anger, the internal combustion engine of the mind. Entranced by fire's finest emblem of the mind—ideas catch like wildfire—fire is our first fascinator and possibly our last, in this a Promethean age.

Part of the appeal of progress is simply the attraction of change, that vibrant principle. (What the caterpillar calls the end of the world, we call the butterfly.) Humor depends on alteration, on the unexpected rhythm of slow slow quick quick change. Heraclitus (was he always stoned?) said: all is flux, all is change. The Yakut shaman of Siberia wears feather-covered costumes which symbolize changing into a bird. The power of all shape-shifting and the enchantment of Ovid's *Metamorphoses* is in the magic of change. And change is the magic in folktales—pumpkins *cum* stagecoaches, frogs *cum* princes. In the animal kingdom, change flips the tadpole into the frog, change is what turns the nymph to the dragonfly. And change is what motivates all the toady, slimy, burpy, throaty things, chortling and gurgling in the mud of amphibibulous delight.

For the North American Algonquin peoples, as for many native peoples, change and exchange is the fundamental process of life, with the exchange of gifts, the give and take reciprocity even between earth and sky, between rivers and rains. To receive means you must give and to give you must receive; all is change.

Changing life-stages are mapped in most cultures; in Hindu tradition, certain ages have certain tasks—including the student, the householder and, later, the spiritual seeker—and it is understood that the individual changes while their world stays the same. In this era as no other, the world changes faster than the individual, which puts unparalleled strain on the nervous system, the result of which, as Alvin Toffler portrayed so dramatically in the opening of *Future Shock*, is that an eleven-year-old boy can die of old age. This is the age of the Temporary, where everything from relationships to architecture, from homes and jobs to art is characterized by fragile transience. Even language is on the move so fast that I must ask a twelve-year-old friend for the meanings of some words. (He never needs to ask me. You see I still say "Oh wick*id*," when I mean "that's great," and that, to him, says it all.) Western homes are transient and the more economically "progressed" the country, the higher proportion of homes are rented than owned. Germany, for instance, has one of the greatest such proportions in the world. It's a way of thinking, this, for as progress has always meant a leaving of place, so in this transience is a constant parting, a mindscape of makeshift, the lean-to of the soul.

.   .   .

Today, there are assumed to be broadly two sides in the debate over society's progress into the future. Those pro-progress, goes the assumption, look forward to the future and include multinationals, cyberspace devotees, space scientists, geneticists, or the nuclear industry. Those agin' it are assumed to be looking backwards to the past, such as environmentalists and advocates of sustainability. ("Sustainability" asserts that all generations have equal rights in their time, so no one generation should rob following generations or impose risks on the future.)

The trouble with this is that the terms of the debate, the either/or choice and the very model of time it uses, linear and arrow-like, are set according to a highly ideological pattern. It pretends to neutrality, but is, even in its language, a heavily weighted argument.

Progressing into the future appeals because it claims an optimistic mobility while the whole idea of sustainability can be characterized as stasis. The English language in general offers rich privilege to terms of movement and dynamics, while many words of stillness and stasis are negatively nuanced ("backward" means thumpingly stupid; "stagnant" is a word that even *smells* bad, like stank or dank) thus fostering the belief that there is somehow an inherent value in "forward" or "progress" and an inherent recoil at "sustaining." Not all languages do this. Classical Sanskrit prefers stiller relationships so most nouns with static characteristics have positive connotations while dynamic nouns of movement and change have a negative value. Balinese, Javanese and Indonesian are all tenseless languages and any time-references have to be deliberately added. Javanese discourages thinking of time as linear or progressive altogether: if two sentences are spoken, "I see a tax collector" and "I run away," it does not imply that the second event happened after, or because of, the first; the running away could have happened *before* the seeing of the tax collector.

"Nothing's happening." In English, this phrase often implies boredom and negative stasis. In Latin American Spanish, by contrast, *"no pasa nada"* is something of a state of grace. Nothing's wrong. (In question form, the two can be closer: to ask "what happened?" in English does imply a concern over bad events.)

In English, then, the very word "progress" has undeserved positive associations; similarly the word "sustainability," time self-renewing and cyclic, has equally undeserved negative associations. "Sustainability" has all the dirgey effort of a worthy cause and none of the dynamite of either "progress" or "change." Who wouldn't rather spend eternity playing with matches among fizzy little devils than spend it humming "Amazing Grace" to a ground bass for the next ten thousand years? "Progress" seems to conjure the vigor of motion, the fire-fare-forwards of life itself. "Sustainability" seems to weigh in with the burden of a heavy stasis, a life half-lived and a death half-died. The opposite is true. Progress, along the trajectory Western society is now on, is a one-word lie; it is neither the travel nor the arrival, but the ultimate ending; not the flame of thought, but a bonfire of humanity. Both sustainability and progress need to be redefined and reclaimed. The fire of the human mind *is* needed, for fireworks and for warmth, not arson; for creating real progress in pyramids of magnificence; for hotly protecting the myriad splendor of diverse cultures and species, in an august roar to increasing, noisy, swampy vitality. Sustainability needs to be re-described, as fleet, hopeful, sensitive and passionate—a vision of the future which tries to look ahead hundreds of thousands of years. Sustainability is not an issue but a synonym. For life. To life, *lochaiyem.*

If there are deceptions in the language, there are deceptions in the very character of the argument. The given choice is either look *forward* to the future as progress (and who could refuse?) or look *backward* as only backward, stagnant idiots do. It is an utterly false choice; believing there are only two choices is putting oneself at the mercy of a mere construct. Someone else's construct, at

· A SIDEWAYS LOOK AT TIME

that. When you are given a choice of only two roads, an old saying goes, take the third. In this case, the third choice is one of neither moving forwards nor backwards but of looking around, not accepting that time need be a straight line at all.

It's hard to do. Western culture is imbued with the idea of time running like a story line from past to future; stories themselves suffuse society's notions of time, intimately connected to the narrative form, be it science fiction, historical fiction or Armageddon fiction, a nifty trilogy of sci-fi, hi-fi and the-end-is-nigh-fi.

Time is often described as tripartite narrative, having a beginning, middle and end. Myths have a Golden, Silver and Bronze Age. History has its Ancient, Medieval and Modern periods; a Stone, Iron and Bronze Age. Marx divided history into an age of primitive communism, followed by one of class society and then a communist millennium. Marshall McLuhan defined three ages, one of oral traditions, then an age of printing and finally an age of electronic communication.

Frank Kermode in *The Sense of an Ending* argues that "tick-tock" is the model of plot, every tick a wee enough genesis, every tock a small apocalypse, but, no matter how humble, the tick will always conjure expectations of the tock. Which is why, he concludes: "time is not free, it is the slave of a mythical end."

The very fact of a beginning implies the need for an end, so from the Alpha of opening, the story must coax an end inevitable as Omega. But does the sense of endings in stories create copycat endings in life? (If *Genesis* opens the book, *Revelation* must inexorably close it.) Ronald Reagan, who notoriously called the Soviet Union an "evil empire," launched his "Star Wars" Strategic Defense Initiative and by so doing perniciously appealed to a national psyche brought up on the narrative of Christ's almighty battle against Satan's armies. Missionaries have been so successful in parts of West Papua that whole villages of indigenous people live in fear: "Christianity has truncated their vision for the future, for they expect the apocalypse and second coming soon, certainly within their lifetime," writes Tim Flannery in *Throwim Way Leg.*

But what if the Bible story had started with plagues and blood-red moons and finished with a sunny garden and damn fine sex? For time is enslaved only by our own structures. Linear stories which must finish; which must seek their ends; and particularly these arrow-like views of time which must reach their target of annihilation, are damaging images. Would society be different if its profoundest models of time were not structured in a past-to-future narrative at all, but if time were seen as an *unarrowed* thing—if, for example, the Bible began with the rhapsody of a psalm and ended with the sashay of the Song of Solomon?

Some of the fiercest proponents of progress at any cost are those who believe that science is value-free. Science, like flame, they argue, has no moral shadow, Prometheus must remain ever unbound. Shooting an arrow of hypothesis, their minds are in flight—this is the breathless trajectory of genius in motion, the dislocated genius which rises above mere morality—for it is the scorching flame of thought itself, *non-pareil* and beyond peril, beyond the good, beyond the evil. Thus the arguments of the nuclear and biotechnology industries.

Lewis Mumford, great analyst of progress, fears one thing above all: "ungoverned creativity in science and invention has reinforced unconscious demonic drives that have placed our whole civilization in a state of perilous unbalance." George Steiner, who so understands the emotional appeal of scientific progress ("For the scientist, time and the light lie before") and a man with no fear of the new, pleads caution now as never before; genetic research, he says, may bring "moral, political and psychological consequences we are unable to cope with. It may be that the truths which lie ahead wait in ambush for man." Science without morality means science without society and its isolate character is mirrored in the intellectual isolation of single-discipline thinking. Although this may have served society well at times, it is, today, an insufficient and divided study. Wisdom, by contrast, demands more connection. And more beauty. There has to be a new ambition for intellectual progress which comprehends science *and* ethics and psychology, which combines knowledge with morality, acuity with kindness, intelligence with elegance, politics with politesse. (Gaia theory, incidentally, according to James Lovelock, "could never arise in the separated and isolated buildings of a university where biologists, geologists and climatologists are tribes apart.")

Science, in any case, never has been value-free. Modern science was engendered in the Enlightenment and imbued with notions of nationalism, the nation-state and of profound misogyny. The *Selfish Gene* theory is Thatcherite, overappreciating competition, underappreciating cooperation. Today, scientific funding is notoriously politicized; funding rockets sky-high for space projects and weaponry which will promote a nation as a political and military entity, not, for example, as an environmental entity.

But at heart, the argument for the unchecked progress of science is an aesthetic argument, working on the same paradigm as the pornography debate. On the one side is the sapphire desire for free speech, brilliant and intoxicating. On the other is the muted beige of complaining indignity, the dull and dogged voice of the humiliated, the solicited. Those who wish to curb the progress of science (ethereal, euphoric and sky-high), those who wish science to work at the behest of morality, take a position which seems tethered, muddy, ground-down and boring. And they are termed Luddite.

The Luddites are one of the most misrepresented groups ever. In the early nineteenth century, they wrecked factory machinery which was destroying their livelihoods and communities. They were not opposed to all technology but specifically to "machinery hurtful to the Commonality." They were not against progress at all, but violently disagreed over what constituted genuine progress. What they sought was democratic progress, for the public good, not progress solely for the rich at the expense of the poor. Today, arise Ned Ludds across the world, dishonestly dubbed enemies of progress, but actually arguing for genuine progress, for the whole of society. Kirkpatrick Sale, in *Rebels Against the Future* (1995), speaks for Luddites ancient and modern. "The pace and range of the technosphere is unstoppable, as if it had a will of its own that no form of public protest or restrictive rule or moral caveat could appreciably affect."

Chellis Glendinning, a New Mexico psychologist, in her *Notes toward a Neo-Luddite Manifesto*, commented: "Neo-Luddites have the courage to gaze at the full catastrophe of our century," which is that "the technologies created and disseminated by modern Western societies are out of control and desecrating the fragile fabric of life on earth." Some of today's foremost environmentalists are happy to be called Neo-Luddites and as technological progress has always been bought at the expense of earthy place, it is appropriate that those muddy protectors of the landscape, the filthy, furious and funny EarthFirst!ers should have their books printed by—Ned Ludd books.

Many people today question the off-ground dislocation of progress, its hostility to place and its character of fire. What they offer instead is a sort of earth-sensibility, in the politics of ecologists, in the work of some artists, in subtle shifts of attitudes to the idea of "home," and the philosophies of indigenous peoples, increasingly respected in the West.

Half a century ago, the Native American Luther Standing

Bear wrote: "We are of the soil and the soil is of us . . . Believing so, there was in our hearts a great peace and welling kindness for all living, growing things." Native peoples would never have had to invent the term "biophilia," coined by Edward Wilson and meaning "the innately emotional affiliation of human beings to other living things." Luther Standing Bear shares his sensibility with the chief of the Guaraní-Kaiowa tribe rubbing earth on his arm and mine. This same sensibility motivates organizations in Britain to title themselves Friends of the *Earth,* the *Soil* Association and Common *Ground.* This same sensibility motivates the philosophy of Wendell Berry, farmer, writer and Earl of stay-puttedness, who lives in Henry County, Kentucky, and farms his land where generations of Berrys have lived; insisting on the importance of place.

This same sensibility gives depth to the work of American artists featured in Alan Gussow's book *A Sense of Place,* for instance, or British artist Robert Maclaurin, his work making of the earth a grandeur which awes, or the work of the creators of Land Art, who go beyond landscape to earth itself; David Nash and Andy Goldsworthy above all, his work giving eloquence to soil, to ice, to leaf. It is the same earthy sensibility which has led to a North American movement towards "reinhabitation"— which author David Abram describes as people "apprenticing themselves to their particular places, to the ecological regions they inhabit." The simplicity of that idea, in today's complex society, is poignant. It means looking neither back nor forwards, but around, looking deep, seeing a mangrove swamp or steppes, meadows, hedgerows or savannahs—seeing place matters.

Real progress needs to be not only democratic in the present as the Luddites argued, but also democratic with respect to the future. Some groups try to think like this—Forum 2000, for instance, which includes Václav Havel, the Dalai Lama, Umberto Eco, Elie Wiesel, and Fritjof Capra among others, is a thinker's parliament set up to consider the future. One of their environmental commissions concluded that there was a certain inevitability for politicians to adopt short-term and environmentally damaging policies as "a result of the time-limits imposed on representative democracy, where electoral terms last only a few years and so to a certain extent restrict the implementation of long-term measures." In London there is a Council for Posterity which has written a Declaration of the Rights of Posterity, aiming in their manifesto for "a planet whose future generations have interests which are represented and protected in the decision-making councils of those alive today." There are calls for a third House of Congress in the States, a "House of Spokespersons for the Future," which would defend their rights. There are two distinct sensitivities in the thinking of groups such as these. First, there is a traditionalism, a sense of past and tendency to conservation. Second, there is a sense of responsibility for, and an imagination about, the future. The coupling of the two is crucial for they combine as an "alliance of the past and the future against the present," as one environmentalist, Victor Anderson, has termed it.

"Future maps" have been published by environmental groups to show how countries will be redrawn as a result of rising sea levels caused by global warming. On some of these maps, island nations such as Tonga in the Pacific and the Maldives in the Indian Ocean have sunk beneath the sea while some fifteen percent of Bangladesh has disappeared.

The World Future Society, based in Maryland, hosts international general assemblies of Futurists, who include environmentalists, politicians, businesses, scientists, information technologists and economists, and their supersessions are titled: Econosphere, Politisphere, Biosphere, Futuresphere, Technosphere and Sociosphere.

For some of these Futurists, the future is considered with generosity and respect. For others, such as businesses and space engineers, it is a market to be cornered by skillful positioning. And if those concerned for the future share an earth-sensibility and appreciation of place, so those who look at the future as a place of exploitation seek progress at the *expense* of place, delighting in the placelessness of future-based technologies (space travel, for instance, or cyberspace), which share an odd quality: a preoccupation with "nowhere," a conception of emptiness, a love of space rather than place. The future is always located in the sky and future-oriented technology is off-ground, a lift-off from human habitation altogether.

The trouble with the computer is that there isn't enough *Africa* in it.

—BRIAN ENO

There is no place in cyberspace—there's no Africa there, no mud or beads or wells or such humanity in the very air. There's no India in cyberspace, no jasmine, no gupshop, no sari, no desert. There's no swampy, mucky, messy stuff, no tadpoles, no owls. There's no nature in the synthetic element and the virtual world so despises the real world that it is denoted on the Internet by the snubbing abbreviation "RL"—"Real Life." Just as roads, symbols of progress, have the effect of blurring place-distinctiveness and homogenizing cultural diversity, so on the information super-highway, particularity is lost, we're on the road to Erewhon, for cyberspace, being a virtual everywhere is an actual nowhere. It is a Teflon place, wiped clean of muddy, earthy reality.

In *War of the Worlds*, Mark Slouka speaks of the cyberspace "technoevangelists" who long for a world where real place and nature is destroyed—what one technoevangelist calls "the irrational, hideous world of trees, birds and animals." Slouka argues that totalitarians and fascists of all colors, among whom he classes technoevangelists, hate mud, hate earth, "have an aversion to the world in all its quotidian messiness."

The Internet is a brilliant tool, for research, for communication. It is gloriously, anarchically uncensored, its expletives deliriously undeleted. It is, unlike so many forms of progress, unusually democratic and can be protectively anonymous, especially in countries without freedom of speech, as internet communications are untraceable.

But. The creation of cyberspace means "cybertime," virtual time which is itself like a fifth element, of Teflon time; within that element, people virtually shop, set up business and virtually fall in love while outside it they would virtually cease to virtually be. (Virtual time is bought with real enough money—the digital revolution has a potential market of $3.5 trillion.)

Manuel Castells, in his exhaustive three-volume work *The Information Age*, describes the network society and information technology creating "a timeless time," negating past and future in a hypertext communicated anytime anywhere. Castells describes the time of industrialism, the schedulization of human behavior, as "clock time" and contrasts both timeless time and clock-time with *"glacial time"*—a key feature of environmentalist thinking; the long-term and evolutionary sense of time, both looking far back into the past and far forward into the future.

But in this timeless-time, or replica-time, it is only replicas of ourselves who can live; hoping to cheat nature, it can actually cheat life. To Internet devotees, the "hive mind" has a powerful lure, progress, as always, seeming to mirror the motions of mind itself. Appropriate to its *meta* qualities, the metaphysical metaplace of cyberspace, this hive mind is *Metaman*. And Metaman is, they say, the next great stage in evolutionary progress. But Metaman is Noman, Nowoman either, individuality lost.

Every cliché leaves you lonely, but Internet relationships have the archcliché, the icon, the loneliest language-sign ever invented.

You cannot have humanity in this timeless zone because relationships are suffused with time. How quickly do you blush; do you pause before you speak; do you laugh before, after or during your own jokes? How soon do you smile after you weep; which is the last gap before you gasp with delight; and is your hesitancy that of sleepiness or shyness? Can you bite your virtual tongue? Is there such a thing as virtual spontaneity? Can you still be a life-long hypochondriac if you're visiting a virtual doctor? If you can have virtual shopping, can you have virtual stealth for virtual shoplifting? *L'esprit d'escalier*, the exquisite and particular sense of lost opportunity, cannot translate into Teflon-time; one unique moment is never gone, because it has never been. Every act in virtual time is final, finite, and finished. No human act is. This is its pity—the act without consequence, cruelty without remorse, sex without hands and jokes without laughter.

The Teflon effect is everywhere in society's views of the future. The multinational oil company Shell has a whole department devoted to "futures planning," where the men in Shellsuits have scripted two "stories," scenarios of how the future could be. *Barricades* is the pessimistic one of the two, describing an "increasingly divided world of rich against poor," "neglected problems worsening," and speaking fearfully of "worldwide revolutionary change."

Their alternative is called *New Frontiers*, a glossy story of "an interconnected world," an intermingling of international interneighbors, "with heightened interconnection and interdependence." The prefix "inter" in Shellspeak is a Teflon veneer of empty optimism; it is brochure-speak, there is no evidence of any genuine conception of the future. Non-sticky and incomplete. Humanity's mixed muck and magic is missing, wiped clean with interrainbowed interniceness.

Shell's efforts to change to new energy technologies are slickly superficial and delicately deferred. "Don't put off till tomorrow what you can put off till the day after" would be a motto for Shell as for wider society. Shell's future has that sleek quality of Elsewhere which so characterizes all futurism. The City of God is always Elsewhere. Cyberspace is Elsewhere, Erewhon is no-place. Witness the cartoon morality of wasting the earth and skidaddling off in a global getaway car to find a new—Elsewhere—home on Mars, in Teflon-time. (Teflon itself is, of course, a by-product of the space industry.) It matters, for, by "dislocating" tomorrow, giving it that singularly placeless character, responsibility for it is diminished, easily rinsed off. In Teflon-time, tomorrow will not stick to today.

Spaceshots only succeed because Americans use the grease of the crushed poor to power them. This thought is "common knowledge" in South America. White people came and took the oil and grease of natives, says an Ashaninca man to me, when we meet in a town in the Peruvian Amazon. "They take indigenous people and cut off their hands and feet, put hooks through their bodies and heat them until the fat drips off. The grease of indigenous people has the most force or strength: it's the best gasoline there is and it is used for planes and cars and to build bridges. It's worth a lot of dollars." Anthropologist Nigel Barley explains in *Dancing on the Grave* how it was told to him: "From the ritual murder of the Peruvian poor and the processing of their bodies, the Americans extracted the 'grease' that is essential for metallurgy, pharmaceuticals and the lubrication of the moon rockets."

This is as literally untrue as it is figuratively appropriate; Western progress has always been bought at the expense of other people. Today, spaceshots are metaphorically bought through the crushing of the poor, as the money for the multi-trillion-dollar space industry is made available by refusing to spend money on welfare and health, and by squeezing money out of countries of the South in debt repayments. Anthropologist Philippe Descola in *The Spears of Twilight* says there is an old Andean belief that "attributes to some perverted Whites an insatiable desire for the fat of natives" which "lubricates and fuels the gigantic machines thanks to which the Whites have imposed their dominion over the world . . . this metaphor of rapacity that has become progressively more literal over the years."

With the affinity progress always has with flame, space progress gains movement through fire, using ion thrusters, its spacecraft firing xenon plasma. Dead guru of LSD, Timothy Leary, was shot into space and incinerated there. (For $4,800, they will launch a seven-gram sample of your ashes on a *Pegasus* or *Taurus* launcher— giving the phrase "shooting up" a whole new resonance—and you circle the planet every ninety minutes for several years, then fall back to earth, being finally reburnt on your re-entry.)

In the name of progress, space joyrides for tourists are planned, Pepsi and Coke slogans are in orbit on the Mir space station; corporate sponsors of space missions want to use lasers to put advertisements on the face of the moon; and there are planned quarries on the moon and Mars. *Artemis* and *LunaCorp* plan to exploit mineral resources on the moon and lunar factories are planned. The far side of the moon, for millennia the mind's most magnificent imagining, has been optioned as a good place for nuclear waste.

And today we have nuking of Mars.

Fire-scientists are considering "terraforming" (the engineering of a planet to make it capable of supporting life) using nuclear bombs, greenhouse gases and solar heat. They suggest setting off thermonuclear explosions below the surface of Mars; using greenhouse gases to heat Mars or putting massive mirrors behind the planet to reflect sunlight back onto it. Some terraformers suggest **273** capturing and diverting asteroids to smash into the planet, generating heat from the crash. Princeton physicist Freeman Dyson thinks a controlled Armageddon is a good idea—breaking up one of Saturn's icy moons and sending its fragments crashing into Mars. The intriguing thing, of course, is why the same way of thinking which wishes to engineer one planet to support life unnaturally should also, using exactly the same tools of fire, work to engineer another planet—ours, actually, our home—into an *inability* to support life naturally. Never mind. There are ambitions to plant Mars with a sort of fake, synthetic life. Onward and upward. *Per ardua ad astro*turf.

Meet Rick Tumlinson. He is the President of the Space Frontier Foundation which staged a space conference in Los Angeles. Here is a man with a space-age vision of progress; he is proposing building something called Alpha Town in space. And he wants to use it to change everybody's thinking. Tumlinson truly understands how the metaphysical appeal of progress is essentially the movement of thought itself. "The Alpha Town concept is not about hardware. It's about a mind-set. It is not a destination or a facility, it is an intellectual framework upon which we build the dreams of human space settlement, instead of a boring Antarctic-style facility on the edge of nowhere, going nowhere." (This is where he is so right and so wrong. Space travel is not a higher level of thinking, but a crude model of thought. The Mir space station is a mere toy compared to the real thing; the human mind.)

But what exactly is Tumlinson planning? What marvel in space is Alpha Town? What architecture of pure imagination? It is *"a kind of industrial estate and dormitory suburb"* growing up around the planned International Space Station. An eight-word definition of modern boredom, not hell on earth but hell off earth, hung from the intergalactic car park at the edge of Erewhon. There is something woefully pedestrian about this mind-set, something turgidly suburban in its lack of imagination. It is as if in this area of human endeavor—which pretends to such sky-rending imagination—the grander the technology, the meaner the mind.

Tumlinson is not alone. Much attention of futurists is focused on nothing more breathtaking than the domestic house: personal diagnostic machines which will tell you what you should eat; smart fridges receiving automatic deliveries from supermarkets; kitchen robots moving food from said smart fridges to smart microwaves. Newspapers relish the attention NASA gives to food in space, printing recipes for food on Mars or the moon, solar-powered celery sticks or hydroponic lettuce. The Japanese Rocket Society, a research group for space tourism, has a brochure full of sickly pictures of a family on space holiday; Mummy looking after Kiddy, Mummy watching Kiddy swimming, Mummy inside ickle round window waving bye-bye to Daddy going *boiing* on a gravity-less bunjee jump in space. And the Hilton hotel chain, we are told, has plans to build a hotel on the moon.

A recent discovery of frozen water on the moon brought two particular responses from space scientists. The first, in the dreary tradition of space-thinking, was that this discovery makes the moon "a kind of petrol station on the road towards deeper exploration of space," as a front page newspaper panted, continuing by quoting a space scientist: "For the first time, we can go to a planetary body and we can fuel up." The second response was a baying for the "colonization" of the moon: scientists and politicians talking of new horizons of progress, of settlers and space frontiersmen.

In one of the most bizarre ideas (supposedly to ease global warming caused by too many streetlights) there are plans to put up "brand new moons" in the sky, metallic mirrors between ten and a hundred times as bright as the full moon. These mirrors will be tilted at specific angles to reflect sunlight onto the earth at major cities in the Northen hemisphere (including Kiev, Brussels,

London, Quebec and Seattle) and are intended to make street-lights obsolete. (The giant mirrors are planned to be attached to a Russian space tug detached from the Mir space station: the name of the tug is *Progress.*) City dwellers who see precious few stars to-day would see far, far fewer. The moon, in a cruel poetic conceit, would be outshone. There were plans to use such mirrors to reflect light over the Arctic, to give it twenty-four-hour, three-hundred-and-sixty-five-day sunshine. The intention of the company doing this, the Space Regatta Consortium, led by the Russian company Energia, is in essence to abolish the night. Wildlife, evolved over millennia to be exquisitely sensitive to patterns of light and darkness will have its patterns of feeding, hibernation, migration and sex disrupted. People in cities in the Arctic region would never again see a star or the moon. Seeds need darkness to germinate. Pips need times of night as well as times of light.

And no one dares say "Get your hands off, it's not yours." British environmentalist Paul Evans comments: "We have yet to fear for space. Perhaps it's time for environmentalists to form a Friends of the Moon." Progress, as ever, is highly political and all the same historical patterns on earth look set to be repeated on the moon, from the vaunted progress of enclosures to the geno-cidal progress of imperialism or modern capitalism, all of which steal what was common and give it to the few.

The moon is yours. And it belongs to the U'wa child and to the Hopi Indians, to the Inuit and to the Mexican woman collecting water and to the monks of Tibet. It belongs to every moth that flies towards it, to every dog who barks for it and to every cat on the hackled prowl for it. Untouched, it is a global commons unparalleled, for it belongs to the *minds* of millions upon millions of people; an emblem of purity, of women, of imagination or of time itself; a beautiful place of thought which can mean a million different things to a million different minds, all on one night. "The moon shines bright. In such a night," as Shakespeare said, it shone for Dido, for Medea, for Troilus and Cressida all alike. But a few hundred overprivileged white males plan probes and moon maps and say the far side of the moon would be a site for an Observatory. Like an exact parallel with "progress" on earth, it uses all the tools of imperialism—from Observatories, maps, enclosures and sheer political force—to colonize a commons. Their progress is not our progress. To lose the moon and be given instead a gas station, a Hilton hotel, an industrial estate, a quarry and a nuclear waste deposit is not a good swap but a terrible theft; a theft from the very soul of humanity. And we, the millions who have lost something so infinitely precious, are expected to *applaud* it?

Goethe's *Faust*, according to critic Marshall Berman, is a narrative emblem of modernity. Faust (another fire-scientist if ever there was one, crackling up with Mephistophelianism) is pictured in his furious progress away from the village of his birth, rushing to build a new home somewhere else. But to create the new means to destroy the old and when Faust—older, wiser, nicer Faust—returns to his childhood's village, his heart is broken by nostalgia for the very world he destroyed, he grieves at the exquisite beauty he was juvenilely blind to before. ("Learn from me," says Frankenstein, "how much happier that man is who believes his native town to be the world." "All human troubles arise from an unwillingness to stay where we were born," said Pascal.)

If Faust is a narrative emblem, Henry Ford is an actual emblem of modernity. Prime-mover in the progress of cars which destroy places and pasts; responsible for the speeded-up mono-jobbing assembly lines which so influenced the time of modernity; his story has a strange end. As an old man, Ford became passionate about the past. He built museums of horse-drawn carriages, sleighs and plows. When he bought an inn in Massachusetts, he ordered the new road to be taken up and replaced with a simple dirt track. He rebuilt his father's farm as it had been when it was his boyhood home.

Modernity's so-called progress is the same journey taken over and over again, progress gained at the expense not just of any place, but of a paradise. And of a home. The precious ordinary. The sedimented especial. The sweet, sweet surprise of the known familiar. The intimately dwelled in and unartfully loved, home.

Astronauts repeatedly see that vision; the beauty of the earth and its unutterable value; what is important is not the creation of faintly ridiculous second homes in the space-age future but the heartfelt appreciation of our first home in the present. Yuri Gagarin's first feeling in space was to see how beautiful was the earth. American astronaut Edgar Mitchell described watching an "earthrise":

> Suddenly from behind the rim of the moon, in long, slow-motion moments of immense majesty, there emerges a sparkling blue and white jewel, a light, delicate sky-blue sphere laced with slowly swirling veils of white, rising gradually like a small pearl in a thick sea of black mystery. It takes more than a moment to fully realize that this is the Earth . . . home.

Aleksei Leonov, the Russian cosmonaut, said: "The Earth was small, light blue and so touchingly alone . . . our home that must be defended like a holy relic. The Earth was absolutely round. I believe I never knew what the word round meant, until I saw the Earth from space."

Since Yuri Gagarin's space flight in 1961, people have burned more of this round perexquisite sphere, the earth, than during the whole of human history before that, for the linear time of so-called progress, that perfect round of cyclic time is shattered. This, as we do not realize, is not the height of imagination but the limit of it, this is not our progress but our exile.

Indigenous peoples have a little joke at the expense of NASA; they have, they say, been going to the moon for thousands of years, in shamanistic trance and ritual, from the shamans of North and Central Asia and South America to the Inuit shamans of Baffin Island (as Mircea Eliade lists in *Shamanism*). And they can laugh at NASA's literal expenses. It costs the West trillions. Shamans do it for free—have mushrooms, will travel. Timothy Leary, Western shaman, should have known better, for he was in orbit long before he died. Pause. Put on that old Beatles LP. All sing. Leary in the Sky with Diamonds.

The Beatles wrote that sky-dreaming in 1967. Lewis Mumford, that same year, published the most respected and insightful history of progress and technology, *The Myth of the Machine*, which concludes that "man has formed a curiously distorted picture of himself, by interpreting his early history in terms of his present interests in making machines and conquering nature." Not tools and technics, he says, but "ritual and language and social organization were probably man's most important artifacts, to give form to the human self using the only tools that could be constructed out of the resources provided by his own body: dreams, images and sounds." Mankind, says Mumford, took so long to work out "tools" because we were concentrating on "the greatest of all utilities first," and that is language. "The pursuit of significance crowns every other human achievement." The real progress is in the mind, be it in shamanistic vision or in the magic of metaphor and the miracle of language, so effably and so ineffably magnificent. Mary Shelley's gentle-hearted so-called monster (who, incidentally, is *much* nicer than Dr. Frankenstein his maker, just much more unhappy) says his greatest discovery was finding that people spoke language. "*This* was indeed a godlike science," the monster pointedly tells his creator-scientist.

And it is to language that modernity must listen now. To find out how fares *real* progress, the progress of the human mind, ask how fare its languages, its diverse and splendid, profligate, myriad-minded expressions on earth.

Not well.

Between five and ten percent of the world's six-thousand-odd languages will become extinct within the next century, say some language experts. Others (such as Michael Krauss, director of the Alaskan Native Language Center) say ninety percent of today's languages will be doomed within a century. Between twenty and fifty percent of existing languages are no longer being learned by children.

Some 2,090,000,000 people (well over a third of the world population) are routinely exposed to English and, David Crystal and other linguistics scholars believe, in a few centuries' time the world may be a more or less monoglot place. This would be, he says, "the greatest intellectual disaster the planet has ever known." Other language experts say the position of English could be taken by Chinese, Hindi, Spanish or Arabic, but they agree that the present wealth of languages will be lost; if not a monoglot place, it will be an "oligoglot" world.

A large proportion of endangered languages are spoken in one of the most rapidly deforested areas—the Asia–Pacific region. When a language dies, a leaf falls from the tree of the human mind. The forests are ancient libraries; the botanical knowledge they contain is enormous, but they are being burned far faster than they can be read. When forests are lost and species are made extinct, biodiversity is threatened and the health of the ecosystem undermined. Working on the same paradigm, linguistic diversity matters immensely, for the variety of thoughts thinkable depends on the language available to express them; the health of the human mind depends on as wide a variety of languages as possible. To lose linguistic biodiversity is to lose untold ways of thinking and varieties of thought; to lose biodiversity of the mind.

This matters for there is a security in having many languages.

If there is only one language and that one becomes warped and damaged, then the human mind itself will be demeaned. If that language is oversimplified for expedience, thought itself will become less subtle.

Across the world, some five thousand indigenous groups consider their languages, cultures, lands and lives are under threat from modernity. Considering that each indigenous group has, apart from a language of its own, also a whole knowledge system of its own, a site-specific science, a medical lore and a philosophy, it is arguable that now, for the first time ever in human history, the sum of human knowledge may actually be decreasing.

From the Chinese beating the Tibetans for speaking their language, the English trying to destroy Irish and Welsh and the New Zealand settlers whipping the Maori language out of the Maori child, language dominance has always been an exercise in political power. But further, the type of language which is dominant reflects the ruling mind-set. Today, it is the English of computer programs, computing-English, *computinglish*; oversimplified for efficiency's sake, its meanings decreed from above, unable to tolerate dissent or plurality. Its language based on commands, computinglish accustoms users to the world of the imperative—program and be programmed, obedience is all—overweighting command structures and undervaluing language's playful, seductive and gainsaying subtleties, its ambiguities and nuances, disagreements and disobediences.

Natural languages are full of time, in their tenses and etymology, computinglish is an ever-present present. Natural languages *live* in their gluey, sticky way of joining word to meaning over time enrichingly. Computinglish uses Teflon, no *dot* word *dot* sticks *dot* to no other. Natural languages have an intrinsic lifefulness because they have a time-resonance but computinglish is hyperdead because it has never lived—it is an artificial intercourse. The dominant language of modernity is a *dead* language and this influences society's deepest thought-systems, making them less compatible with nature than with artifice, less at ease with life than with death. Living languages are self-propagating, spawning seed syllables, stretching ambiguously, syntactically tactile, diffuse and suggestive as sex itself, abundant, panglossolalic, profligate, vocative, evocative and provocative. Now slippery with fictive fascination; now stiff with austere truth; now tickled with trenchant absurdity; language luxuriates in a careless, splendid, inefficient uproar *I AM ALIVE*. Computinglish is a language sterilized and wantless as a hospital death.

Beyond language-dominance and computer-English, how else are the ideas of "progress" and "language" related? Language directly depends on nature, on the sounds of the forests and the calls of the birds. Language began, suggest linguists, with mankind copying animals, as David Abram comments in *The Spell of the Sensuous*: words came hooting like owls and barking like dogs, murmuring like the wind and cawing like the crows. No hunting society could afford not to understand and imitate the calls of the animals it hunted. Indigenous cultures still maintain this connection: the Swampy Cree of Manitoba say they were given language by animals; for the Inuit and many others, humans and animals once spoke the same language. (Western culture mocks such an idea as the infantilism of a Doctor Doolittle talking to the animals.)

The Koyukon Indians of northwestern Alaska listen to the song of the diver or loon—the loon, lovely bird, like a silky, feathered crossword in flight—to compose their lilting songs and the Koyukon bird-names are onomatopoeic so to name them is to imitate them. The great horned owl the Koyukon call *nodneeya* which means "tells-you-things." Destroying bird habitats brings a silence of birds which is also, subtly, our own silence, as human speech loses one more string in its bow of imitative capability.

Western engines of "progress" destroy natural place—for the

Erewhon of endless suburbs, or anywhere-freeways—and destroy that place's specific language; no man's language is spoken in no man's land. By contrast, when cultures retain the relationship between language and place it enriches both, a land alive and speaking. Apache Indians have place-names grounded in particularity; Big-Cottonwood-Trees-Stand-Spreading-Here-and-There, or Coarse-Textured-Rocks-Lie-Above-In-A-Compact-Cluster. Compare that with Surbiton. To destroy the land is to destroy its language. "Only where there is language is there world," as Adrienne Rich wrote. (*Langscape* is the neat name of the newsletter of Terralingua, an international advocacy organization for linguistic and biological diversity.)

When Western progress destroys landscape, it destroys language as directly as felling trees: to silence the birds is to silence a part of human language and to render a species extinct is to make a simile die. In English, from spitting like a cat, badgering someone, being foxed, howling like a wolf, to leapfrogging, being a lounge lizard, looking owlish or being batty, having doe-eyes, squirrelling something away for later, fighting like a tiger, being a fatcat or a chameleon character, offering a fishy remark or gulling someone, language is steeped in the natural world, so well-versed in nature, so well-read in tooth and claw.

And you can, just, glimpse a simile in its halfway stage—as animals or birds become rarer, their metaphoric incidence decreases, so "to skylark around" was once as common an expression in language as the bird was in the high sky; now it sounds antique because the bird is rare. Thanks to industrialized farming in Britain, three million skylarks have been lost in one generation. The linnet and tree sparrow have declined by more than half in twenty-five years; also endangered are the song thrush, the bullfinch, the corncrake, the turtle dove and the gray partridge. And within one language, dialects are lost, so the bird known as the yaffle, or the rainbird in Cheshire, the heigh-ho in the South of England, the stockeagle in Mid-Wales, and the popinjay in the Middle Ages, is increasingly only known as the green wood-pecker now, the varied words becoming rare. The otter, hardly seen in Britain today, is rare in the language, though it splashes around in Shakespeare, where Falstaff, having served up Mistress Quickly with all manner of similes—stewed prunes and drawn foxes—calls her an otter, neither fish nor flesh.

Shakespeare's locale teemed with species now rare and, simi-larly, his vocabulary teems with the variety of the natural world. (His writings, appropriately, have been described as being "like a work of nature.") Take one scene at the Boar's Head Tavern, where Falstaff (he of the great lines: "then am I a shotten her-ring . . . if I fought not with fifty of them I am a bunch of radish") says: "I am withered like an old apple-john . . . I am a peppercorn, a brewer's horse," and "thou hadst been an *ignis fatuus*, or a ball of wildfire . . . an everlasting bonfire-light, . . . that salamander of yours . . . Dame Partlet the hen . . . foxes, ot-ters, lion's whelp." Falstaff, swag-bellied and bountiful of natural imagery, closes: "Rare words! Brave world! Hostess, my breakfast, come! O, I could wish this tavern were my drum," banging on gloriously like a one-man-biodiversity band, the periabundant vi-tality of periwinkle and papaya, this pouring language, raining radish and reindeer. If Shakespeare himself were alive today, his very language would be sieved by species-loss and rarity to a poorer strain.

The rest is silence. Landscapes are full of noise but when they are destroyed, as a result of progress, there is left the bleak quality of silence, the wild and wide call of the boobook owl gone, the skylark and nightjar (the whippoorwill) silenced. There are oth-ers—the silence of screen languages which blanks out noisy liv-ing languages; a Saro-Wiwa silenced; the silence of Alpha Town hung in silence. Peoples and cultures fall silent, voiceless; the

U'wa who kept the world alive by singing it are silent on a cliff of silent fall; the U'wa birds who create the places they fly over by singing them are killed by a word—progress—and there are no flutings, trumpetings, croakings, chatterings, no coo, no whistle, no riff. Skies empty, silence falls like acid rain.

Picture this. A crowd gathers to watch a rocket of brilliant light arc across the sky. In Egyptian myth, the dying Pharaoh creates future generations, shooting his sperm skywards in an arc of astral projection to impregnate the star-goddess Isis, who bears his son. To the Pharaohs, first cyberspunks, the future was a fertile conception. The ancient Egyptians "conceived" the future in two senses; they imagined it, picturing it as that arc of brilliant light shot across the sky (the sky, as ever, the site of the future), and they also conceived the future sexually; the present generates, conceives, the future through the generation of children.

"We have not inherited the Earth from our parents, we have borrowed it from our children," as the famous saying has it, attributed to Native American Chief Seattle. Many cultures conceive the future and plan for it by looking ahead seven generations; the Iroquois Confederacy of Six Nations, for instance, living in the remains of their ancestral land in America and Canada, consider the effects of every decision they take "unto the seventh generation." African and Polynesian tribes, too, were, traditionally, said to look ahead at least seven generations. Seven generations, it is thought, is chosen because that is the greatest number one could hope to know in one's own life; one's great-grandmother, grandmother, mother, sister, daughter, granddaughter and great-granddaughter.

Western politicians have no such conception; former American Secretary of the Interior James Watt, advocate of short-termism, said that if the end of the world is near, then it makes no sense to save anything for a nonexistent beneficiary—Teflon politics for Teflon times. And, of course, the power of the American government to poison hundreds of generations into the future is far greater than the power of the Iroquois to protect their seven.

Looking ahead seven generations, say a hundred and seventy-five years, seems impossible to modernity, but to many Western leaders, it isn't even an aim. Economists, such as Robert Solow, argue that the present *should* deprive the future of nonrenewable natural resources, for—assuming as he does that nature can be replaced by financial capital—tomorrow will have more capital than today if natural resources are turned into capital resources. John Shanahan, Policy Analyst for Environmental Affairs at the right-wing "Heritage" organization says: "By denying ourselves material wealth today, by slowing the accumulation of wealth, we are denying our children. You deny the future by not using resources now." George W. Bush, honoring short-termism, rips up the Kyoto agreement. For the sake of immediate oil profits, long-term environmental security is jeopardized. "I'm with Bush," comments James Burkholtz in the spoof publication *The Onion*: "I'm with Bush: What have future generations ever done for me?"

Good at Latitude, bad at Longitude: Western society tries to be good at lateral responsibility, to its immediate peers, but is irresponsible about longitudinal responsibility, down the generations. It's similar to Western "time," where the lateral time of the present is ubiquitously overconceived, while the future is barely conceived at all.

In contrast to the rampant chronocentrism of today, seeing the present as far more important than either the future or the past, Zen Buddhists define their task as "infinite gratitude for the past. Infinite service to the present. Infinite responsibility to the future." Edmund Burke, writing about the French revolution, said: "Society is indeed a contract . . . not only between those who are living but between those who are living, those who are dead and those who are to be born." To honor such a contract you need first to conceive of the future.

But *does* modernity really conceive the future? Language doubts the reality of tomorrow. Consider the words for describing society in time; post-industrial, post-modern, post-communist. Where are the *pre* words, pre-contented, pre-chirpy or pre-contentious? Vocabulary, offering postscripts but not prescripts, declines to predict society's future, preferring to *post*pone its judgment. Films conceive the future with fear for humanity; *Blade Runner* in 1982, for instance, with its frightening underworlds, or the "post-human" cyborgs of *Terminator* in 1984.

The Elsewhere quality, the Erewhon character the future has, when people speak of it, tells of the same unreal conception. In the end, for all Shell has its Futures department; for all cyberspace devotees talk cyber-futures; for all the space-industry spokespersons gleam on about Alpha Town and colonizing the moon; for all the World Future Society tries to map it and picture it; there is in all a smooth patina of irrelevance compared to society's actual treatment of the future, the *actions* of modernity reveal a radical inability to conceive it.

Synthetic chemicals and genetically modified organisms are pumped into the environment with no concept of their future effects. Modern lifestyles cause pollution of air and water which will choke the future. People's mental horizons into the future only usually extend a few years, whether you judge by personal schedules (almost no one will make diary arrangements over a year ahead) or by social horizons—as witness the millennium bug which, though knowable decades ago, only came to bother the public imagination about a year before the said millennium. Such horizons are probably "in our genes": humans in a natural environment would hardly need to look ahead beyond the cycle of one year, sowing to harvest, or annual herd migration. The "limited" horizon is only so problematic today because modernity's power to affect the future is far greater than its power to imagine that future.

The wrong things wither, while the wrong things last forever. Time capsules are one popular way for the present to link itself to the future, preserving self-descriptions and self-presentations, in the form of diaries, photos and news cuttings. The Canadian Conservation Institute says that the contents of capsules frequently reveal "a not-altogether-surprising lack of imagination" as these capsules most often contain "documents of purely parochial interest, albeit of a timely nature" and they then—without irony—proceed to advise people to make capsules of calculators and wrist watches and say "another aspect of our throw-away society is packaging material, some of which is very skilfully designed and says much about our attitudes . . ." Indeed. For far outweighing these tacky "presents for the kids" of watches and calculators will be tons and tons of litter, equally solicitously preserved, as litter-engineers design their packaging to last hundreds of years, presents for the future, dodgy cigarette lights, plastic bags, rusty cars and plastic tampon applicators. My my, how kind.

The future is a blank absence of elsewhere: there is a Teflon coating between today and tomorrow. It is an attitude so implicit it is all but invisible and one merely masked by forecasts, plans and futurism. This is never more clearly shown than in the nuclear industry. If society truly conceived of the thousands of years into the future which it is poisoning, would it be able to do it? I try to conceive that length of time. I can't. I've never met anyone who could. Nuclear waste requires a vigilance for thousands of generations into the future which even this present generation cannot muster: witness the Dounreay nuclear plant in Britain, where Polyfilla, plaster of Paris and Pyrex glass were used to encase nuclear waste.

Were the mind-set of the nuclear industry to be described as an individual, it would be called a psychopath, psychopathic in its urge to harm, in its inability to feel anything for its victims and in its murderous lack of imagination. The land poisoned by Chernobyl will be unsafe now for thousands of years, so no child, no chive, no cabbage and no king will grow healthy there. The psychopath merely shrugs. Childhood leukemia clings like an invisible shroud around the U.K.'s Sellafield nuclear power plant, now and forever, now and forever, now and forever. The psychopath smiles thinly. Radioactive nuclear waste is encased in glass or concrete and buried underground or at sea. Glass or concrete, knows the psychopath, will not last as long as the toxic waste. So what? The toxicity of nuclear waste will last many times longer than all recorded history. Who cares? The psychopath will not be there that day when all the radioactive waste will leach out, come stealthy, now, come toxic, to kill the future. The psychopath feels nothing, smiles coldly and stubs out a cigarette on a child's face. Why not?

Modernity's carelessness for the future would also seem to run counter to the nature of biological systems (of which mankind is one) for living organisms, supposedly, are "wired-in" to the future, and they propagate and invest time and energy for future survival. To lose that connection means the species risks extinction.

Scientists at Los Alamos, New Mexico, are working on the Knowledge Preservation Project, to preserve forever the formulae for nuclear weaponry which they see disappearing as a generation of scientists dies. "We don't want to push the erase button on our memory," says one. If only. To remember this costs forty billion dollars. To forget would be priceless.

Demonstrably, society can forget. In contrast to nuclear knowledge, book knowledge has the shelflife of chip wrapper paper. According to Jean Chesneaux, in *Brave Modern World*, "Our books are printed on paper so acidified that it will disintegrate in a few decades. They have little chance of cluttering the shelves of the 21st century bookshops which . . . will continue to offer customers older but more durable volumes manufactured in less callous times." (For exactly this reason my granddaughter, should I have one, may never be able to read these pages. I could bequeath her my bookshelves but not my book.)

Pollution, litter, unpredictable genetically modified organisms, synthetic chemicals and, above all, nuclear waste, are society's truly representative souvenirs and compared to them, the very idea of time capsules' selective memorabilia is a blushing absurdity. The future will need no artificial reminding of us, the evidence will be everywhere, the remains of our housewrecking party, the forest furniture this generation burnt for firewood, coal cellars emptied of fuel and resources, the blocked sink of polluted oceans, the freezer door left open so the ice-box ice caps melt and unexploded nuclear fireworks half buried in the back garden, with the instructions left on the table, *try this yourselves, kids.* Will future generations really ignore all this and take as our last will and testament only the time capsule on the mantelpiece and with milky gurgles of delight seize on a wristwatch and a plastic calculator?

Other societies might see the future as an act of conception; modernity treats it as an act of blind will. The Egyptian image—the white sperm's arc across the heavens—is the arc of imagination and of conception: the nuclear arc which modernity imposes on the skies of the future is the arc of will, the shape of cannon-fire or rocket launch.

October 15, 1997. Picture this. A crowd gathers to watch a rocket of brilliant light arc across the sky. The ancient Egyptians would have fainted with recognition. The space probe, Cassini, is off on a seven-year journey to Saturn. (On board is a little probe called Huygens which is aiming for Titan, one of Saturn's eighteen moons.) Now the Cassini space probe is powered by 72.3 pounds of plutonium, the most toxic substance known. One pound, hypothetically, could give the whole world lung cancer.

To reach Saturn, the rocket must orbit Venus twice and then earth. In August 1999, it was 312 miles above the earth's surface, traveling at 42,000 miles per hour, and within 35.36 seconds of what NASA calls "inadvertent re-entry" which could scatter plutonium across the world. Dr. Ernest Sternglass, professor emeritus of radiological physics at the University of Pittsburgh, said that exposure could mean up to thirty to forty million deaths. And NASA didn't even know the chances of Cassini releasing plutonium; at one time it said the chances were one in five thousand, then one in three hundred and forty-five.

Cassini could have gone up solar-powered, say protesters. Why didn't it? The likeliest answer seems to be that the American government wants to develop nuclear power in space for future nuclear warfare. Lt. Gen James Abrahamson, former head of the Strategic Defense Initiative organization, says: "Failure to develop nuclear power in space could cripple efforts to deploy anti-missile sensors and weapons in orbit."

Ex-President Clinton, who could have stopped the project, supported it. Cassini is the inverse opposite of the Egyptian pharaoh's sperm line: the pharaoh's penis shoots a majestic arc to create the future by conceiving a child; the president's "probe" of a plutonium arc threatens to kill the children of the future. The obvious links between the phallus and the space probe, the rocket and the penis do not need to be repeated; they are carved, now, into the sky itself with a hard *arc de triumph*allic power. The fact that Bill Clinton, "Slick Willy," the Teflon-president for these Teflon-times, has a real penis which makes world headlines for months, while the metaphoric phallus of Cassini hardly raised a murmur, simply reveals modernity's inability to comprehend the magnitude of its own actions and its unceasing appetite for anything which doesn't matter.

And, entirely fitting for an age which finds it so hard imaginatively to conceive the future, modern society is quite literally finding it hard to conceive. There are frightening increases in testicular cancer, sperm counts are declining and many studies reveal abnormal amounts of damaged sperm, thought to be caused by man-made environmental pollutants. A book, *Our Stolen Future* (Colborn, Myers and Dumanoski), details a horrific spread of reproductive problems among humans and other animals due to hormone-disrupting synthetic chemicals. These include, for women, miscarriages and ectopic pregnancies.

The Cassini project is a wealth of cultural allusion about time. The Cassini family were among the foremost astronomers of the seventeenth and eighteenth centuries. The Huygens probe is named after clockmaker extraordinaire Christiaan Huygens. The Cassini space probe is bound for Saturn and the Greeks identified *Kronos*, Saturn, with *Chronos*, Time.

Saturn, of course, is the murderous father of myth, eating his children, gorging himself on the future. Chronos gives his name to Time the devourer rather than time the creator—the chronological clock-time of modernity. While the ancient Egyptian *arc en ciel* was all for sex, for the creation of children, the arc of modernity, epitomized by the Cassini project and the nuclear industry, kills children. In its attitude to its children, to its descendants, to the future, the whole drama of modernity is a grand replay of the myth of Saturn. This individual generation enjoys a greedy lonely glory, as the words attributed to Chief Seattle suggest, speaking of the white man's loneliness of spirit and his greed to consume: "He kidnaps the earth from his children. His appetite will devour the earth and leave behind only a desert."

In 1827, George Cruikshank drew a brilliant cartoon of Chronos, Time the relentless devourer, sitting at a table, eating plates of people, places and elephants, ingesting all before him; Consumerism Unlimited. It is a portrait for today; until now, humankind has been able to think of itself as the child of time. Now, past the critical moment when we started using more than nature could replace, started polluting faster than nature could clean, we have grown up. We have become Saturn, eating our progeny. We have become the very cartoon of creation, picnicking on our own children, not out of anything so grand as fear or self-defense, but out of casual cannibalism, stupid self-indulgence and chronic greed.

"Whatever will be, will be," as the innocent song has it. What "will" be is the future, what simply "will" happen. Stop a moment on that word "will," for it is telling. "Will" is not innocent. What will be is not in the lap of some god-of-the-future, but is an act of will, an act of power, the will of today.

If literature teaches us anything, it is that when the will is infinite in its grasp, the only possible result is tragedy. If history teaches us anything, it is that forcing one's will on another is the nature of cruelty and of war. If humanity teaches us anything, it is that will must be tempered with respect and this is never truer than in our relationship with the future; for whom we make our will. This will could be a present, an act of care and generosity, to make a will as a gift to the next generations. But our last will and testament is a will indeed, the will of the present imposed on the future. It is a legacy in reverse, a demand for payment addressed to the future, a bill presented to the kids who weren't invited to the party. The pharaohs conceived a lustrous shining future for their children, writing their will across the sky in starry seed, a rainbow of imagination arching from earth to sky and back again. These generations, unregenerate, have a squalid fantasy, scribbling our last will and testament across the sky in plutonium and acid rain, our will *will* be done. Willy nilly.

I am at two with nature.

—WOODY ALLEN

It is dawn and in Gisborne, New Zealand, first city in the world to see the day, the sun rises, washing along the lines of the land and on to Australia and up over the island of Mauritius. A little further west, the sun sprawls tropically up the sky over the island of Madagascar. Over Israel and the Middle East, the sun ascends, a call to prayer. It slopes along the rim of Italy, over Virgilian landscapes, and tilts a warm shadow up against a sundial on a farmhouse wall in the South of France. It laps African savannahs and rises over Nigeria at the same time as over Stonehenge, making bars of the stones set up to worship the sun. Then, further west, the sun lifts the lid off what's left of the night in a Dublin pub, scoots across the Atlantic and scrabbles up the sky over the West Indies. A cock crows, someone swears, ups and dresses. (The cock crowing at dawn in the West Indies is said to screech *"Gi' me trousers,"* man's clock chiming with nature's—and cock with cock.)

And and the the sun sun also also rises rises (*sic*) (*sic*) over Dolly the sheep and Dolly the cloned sheep in a laboratory outside Edinburgh.

The sun represents time and nature absolutely equally; it is the prime timepiece of nature, an indivisible unity of time and nature. You cannot split the sunrise from the sun, or the hour from the light, or the dawning from the day. You cannot say whether the sun is primarily time or primarily nature, here the two are one. Like rolled gold, the sun rolls time and nature together, peerlessly unique in its symbolism of both time and nature. In the Mayan language the most common word for time, *kin*, is shown by a hieroglyph meaning "sun" and "day," and likewise in the Karen language, the word for "day" is also the word for "sun." It illustrates their feeling of a constancy between time and nature: an understanding taken further in other societies; some ancient Hindu texts assert that the sun is the source of time.

But what is this relationship between time and nature? Poetry, ritual and language itself all describe it as a marriage of *Father* Time and *Mother* Nature; and this cornucopian copulation of time and nature is perceived as the genesis of life, reproducing from a parental source.

The relationship is the ultimate conceptual passion of a Golden Age; time and nature coming together in a splashy, skirtless extravagance of fecundity, time's vigor coupling with nature's vitality in the warmth of the sun, as Baudelaire wrote:

> *J'aime le souvenir de ces époques nues*
> *Dont le soleil se plaît à dorer les statues.*
> (I love the memory of those naked epochs
> When the sun amused itself with gilding statues.)

It's a common-law marriage for an age of Commons. "Everywhere the commons will breathe of spice and incense," writes Virgil in the *Eclogues*, and thus in the *Georgics*:

> *Before Jove's time, no settlers brought the land under subjection;*
> *Not lawful even to divide the plain with landmarks and*
> *   boundaries:*
> *All produce went to a common pool, and earth unprompted*
> *   was free with all her fruits.*

That nature was bountiful is only part of the story. Arguably, the real bitterness in any nostalgia for the Golden Age is that then *time* was bountiful. Not only nature, but also time was a "commons" unenclosed, unfenced and unfettered; a wilderness of time.

The abundance of nature and its Everywhere-A-Commons feel is exquisitely demonstrated in Virgil's repetition of "here;" everywhere is *hic,* here. "Here among hallowed springs," "Here with me in the woodlands," "Here are two altars . . . here in summer shade or here in winter firelight." As "here" represents the abundance of nature so the reiterated "now" represents the abundance of time, All-Time-A-Commons. "Now let's hear . . . ," "Now fold the flocks." When Mother Nature couples with Father Time, the word on her lips is "Here, here;" "Now, now," the word on his.

When the sun rises over Australia, it rises over the ancient Aboriginal Songlines—perhaps the most sophisticated image of the indivisibility of time and nature. The song, by virtue of its pace, its rhythm and musicality, represents time. The land represents nature. The singer walks and as they walk, the landscape changes. As the landscape changes so the singer is reminded of the next stanzas. The land acts as a mnemonic for the song and the song "creates" the land it evokes. Nature and time are indivisible in their reciprocity. (It is as if the Aboriginal Australians had an intuitive understanding of the concept of a Space-Time continuum—inseparable, observer-dependent—which the Western world only reached painfully, mechanically and recently.) The song enhances place, even creates it, while place, in turn, creates the language of the song. The Ancestors sang events and names into the land as they crossed it and the Songline is the path the Ancestor walked; so when an Aboriginal person today walks that Songline and sings that song, they are part of the re-creation and re-enchantment of the land.

In another sense, they represent both time and nature in their *pace.* American poet Gary Snyder recalls traveling in a pick-up truck with a Pintupi elder in Australia who began rapidly to recite a Dreamtime story about Wallaby people and then—quickly-quickly—one about Lizard girls, telling this story fast, and the next faster until Snyder remarked: "I couldn't keep up. I realized after about half an hour of this that these were tales meant to be told while *walking*, and that I was experiencing a speeded-up version of what might be leisurely told over several days of foot travel."

The Hopi express the unsplittability of time and nature in their very language. Time, to them, is not an abstract mechanical thing, but an intrinsic process of nature. Time happens in the moment of the biting of a scorpion, or in the months of a crocodile growing up, time is identical to the maturing of the baby whip-poorwill. It is the process, or duration of nature. One indigenous tribe in Madagascar, tying time directly to nature, refers to a moment as "in the frying of a locust." The Cross River peoples in Nigeria, describing the exact length of time it took for one man to die, a period we would call less than fifteen minutes, said that "he died in less than the time in which maize is not yet completely roasted." In Tibet, traditionally, it was said that a boy could become a novice monk "when he was old enough to chase a raven." In both Spanish and French, time and nature are connected, as *tiempo* and *temps* mean both time and weather in those languages.

The English language can show time's duration by referring to an act of nature. "While" is particularly fruitful. E.P. Thompson quotes "pissing while" which as he comments is "a somewhat arbitrary measurement." Wycherley writes "Stay but a making Water while (as one may say) and I'll be with you again." Browning in *The Ring and the Book* says, "Be it but a straw 'twixt work and whistling-while." The *Oxford English Dictionary* yields some delicious ways; it records "breathing-while" and "life-while."

You can also find "supper-while" and "alle the durge wylle," "paternoster-while" and "miserere whyle."

In Scotland, the morning's first light was called the *neb,* the beak or nose of the morning, or the *scraigh*—screech—of day, after the shrill cock-crow. Twilight, the uneasiest hour, the neither-nor, was the "gloaming," "between-the-lights," the "mouth of night" or "cockshut." Midnight was called the "howe-dumb-dead of night."

From the sun's daily arc to the year's arch of seasons and the small curl of time in insect life, nature is full of time. Springtime is indicated in hundreds of ways: in the movement of toads; when gorse begins to smell of coconut; when the oak-loving moth, the Spring Usher, is first fluttering; when the yellowhammer's song is first heard, calling for "a little bit of bread and no cheeeeese." A bug-eyed beguiler of the insect tribe knows when to hornswoggle its prey and the bee knows when to hexagonalize, the wildebeest to migrate. The lugworm is time-conscious—feeding with exceedingly regular enthusiasm at three-minute intervals—but not very bright, for it does so whether or not there is actually any food. Potatoes can tell the time, as humans have discovered by measuring their rate of oxygen-use. Even with its eyes gouged out and sealed up in a container of absolutely regular temperature, air pressure, light and humidity, your poor blind potato knows the season of the year and the time of day. In this, the spud is brighter than the human, for we, similarly blindfolded, cannot judge it so accurately. According to an old edition of *The Countryman,* plants such as hops, honeysuckle and convolvulus, originating from countries north of the equator, twist clockwise with the sun. Sweet chestnut seems to do the same. The scarlet runner curls counterclockwise, suggesting it is indigenous to countries south of the equator. And vines, they say, choose. Planted in the north, they twist clockwise, while in the south they turn coun-terclockwise. Sure-yani Poroso says of the Leco people: "We can tell the time from certain flowers which grow on trees and which turn their petals towards the sun as it moves across the sky during the course of the day; so even if it's cloudy you can tell the time, because the flowers know exactly where the sun is."

Chief Standing Bear wrote in *Land of the Spotted Eagle* that "Time of the day was kept by the sun . . . The sun was a very im-portant symbol in the Sun Dance of the Lakotas . . . The sun-flower, which grew in great abundance on the plains before they were upturned by the white man's plow, was the symbol of the sun . . . adored for its golden beauty, remindful of the sun, and because its face was at all times of the day turned toward the sun . . . when the Lakota arose each morning to meet the sun he met a personality—the emissary of Wakan Tanka."

When the sun rises over the South of France, it touches one ancient sundial, on a farmhouse wall in the Cevennes. The clock face, seen from a distance, is a tawny sun, with a round gold cen-ter, and rays spread outward from the pointer's (the gnomon's) base. (The depicted rays of sunlight, pointing to the hours on the clock, are composed of the leaves of the *chardon soleil,* the sun thistle.)

Like this sundial, all the first clocks used nature; sundials needed the sun; water clocks (*clepsydrae*) needed water; candle-clocks needed fire and sandglasses sand. Star-clocks need sky. (A star-clock was used in In Salah in the Sahara; there was a star for every person who wanted to draw well-water, so when that star appeared in the sky, their turn had come—those waiting for their stars were called the children of the stars.)

But there were problems. Sundials were confused by clouds, water clocks stuck if water froze and all "natural" clocks were use-less in the dark. The spice clock, invented, it is said, in seventeenth-century France and designed with spices instead of numbers, so you could taste or smell your way through the hours of the night, provided one ingenious solution.

The invention of the verge escapement with foliot does not, let it be said, seem sizzlingly sensational, on the face of it. Oh, but it is, for this, the invention of mechanical clockwork, early in the fourteenth century, marked a major triumph over nature. With mechanical clocks, artificial time got wheels. It brought in what John Zerzan has called "a qualitatively new era of confinement" as "temporal associations became completely separate from nature."

In the countryside, until the Middle Ages, time was inseparable from nature. Work patterns, the timing of harvest or lambing seasons, all set rural, agrarian rhythms. It was a time, writes medievalist Jacques LeGoff, "free of haste, careless of exactitude, and unconcerned by productivity." Time had immediacy and radiance; it was a sensual perception not a notation. In the growing cities, though, life became very different. Neither work nor social life in the city depended on nature; there time was "constructed." Clockwork replaced natural rhythms. Today, it is all but a cliché to say that time runs more slowly in the countryside than it does in cities; people leave cities for holidays, to go where there is "more" time, because there is less time *measurement*—fewer clocks. In this way, as in so many others, the clock is, in effect, the opposite of time.

Whereas nature is full of time, urban life is full of clocks. Time measurement has deep symbolic resonances with citification; from the town-hall clock to the Romans who counted their history from the foundation of the *city* of Rome. Christianity, whose fight with paganism can be read as the struggle for the city and against nature, found a friend in clockwork. Temporal discipline suited it, both in a practical and in a metaphysical sense.

In Judaism, in ancient and medieval times, nature, not clockwork, decreed the times of the prayers, so the morning's first blessing was a chirpy thank-you to god for giving the cock wit and nous enough to know the difference between night and day. In the evening, prayers began when you could see three stars, clouds permitting. Clouds not permitting, they began when you could no longer tell whether the sky was blue or black. In contrast to the living poetry of this time distinction, Christianity chose the prosaic, advocating the punctuality of prayers according to fixed hours, from the third century onwards.

Some of the steps towards the ever more divided hours and ever increasing precision of today's time can be seen in astronomers' records. The great Danish astronomer Tycho Brahe only twice even mentions minutes in his journals of 1563 to 1570. In 1577, he describes a clock which would show those unremarked minutes, while in 1581 he is speaking of seconds. In the eighteenth century, says historian Anthony F. Aveni in *Empires of Time*, another crucial abstraction followed.

> There is no choice: if you want uniform hours in the day, then you must cast your eyes away from the sun in the sky, and so in the eighteenth century, astronomers banished the *apparent* sun and put in its place the *mean* sun. . . . This fictive sun and its time, which we keep on our machines, gives us the equal hours and days we cannot seem to do without.

Today, that other great timekeeper of the countryside, the cock crowing at dawn, is in trouble. City-dwellers, using city-time, who move to the country have initiated an avalanche of complaints against cocks, including the now-infamous, expensive legal case in Britain attempting the silencing of Corky the Cockerel. Corky has been replaced by the alarm clock, the mechanical rooster.

· · ·

"Industrial Agriculture" is not a phrase to conjure with; the very words would have been a sour oxymoron to Virgil, for when he wrote of agriculture, he explicitly contrasted countryside with city and free Commons with commerce. Farm machinery, chemical fertilizers and assorted pesticides mean that nature's time is now dominated by human—urban—time; crops which naturally flower once are forced to double harvests, artificial fertilizers force-feed the earth to ripen its seed at modernity's speed and seeds are sown according to economic demand. (Ancient beliefs, by contrast, still held by some, speak of planting in time with the moon's phases, for it was believed that crops planted when the moon was waxing grew better and yielded more than crops planted in its waning.) Indigenous peoples of the Peruvian Amazon traditionally did not cut trees to make canoes or build houses when there was a new moon. The moon is considered "green" and at this time wood is also "green." People's bodies too, at a new moon are considered green; tender and exposed. Today, science explains: at a new moon, the sap rises in the trees and the wood is not good to cut.

With industrial agriculture, time is reduced to a sequence of numbers without the vibrancy of natural seasons or lunar processes and nature is forced to work at clockwork pace. The relationship, playfully intimate, of Mother Nature *in flagrante delicto* with Father Time, is cut short, as if society had seen in it some seedy incest. Mankind is here, telling Time to keep his hands to himself, slapping surrounding Nature to stop her breathing her fecund swarmy breath, inspiring the hours with houri's ideas. Divorcing nature and time makes an artifice of time and an artifact of nature, desexing both, to make money out of them: as modernity insists on a commanded time and nature for an all-commanding economy. Time was, time would vary; sometimes basking in the sun, sometimes nervy with a harvest threatened by rain, sometimes swampy, sometimes strict. Now all clocks are cropped to a uniform, military time and *natura naturans* is a perplexed nostalgia, its crops now clocked to uniform, row upon row, now sow now grow, getting the hops to hop to it.

Today, with genetic engineering, time and nature are even further divorced. Many crops are engineered specifically to alter their relation to time, some bred to overcome time's seasons, so cold-resistant genes are introduced into tomatoes from fish in order that they can survive frost. Some crops are engineered in order to fit into the human time-schedule; strawberries are sprayed with Frostban in order to artificially lengthen the growing season; and the sunflower, its golden largesse of full summer sun, so symbolic of nature's time, is modified with genes from the Brazil nut to improve its storage quality.

There are arguments for the genetic engineering of nature. It could lead to a reduction in food chemicals, to higher yield crops and to cheaper medicines. But genetically engineered plants are likely to increase the use of herbicides and pesticides, accelerating the evolution of superweeds and superbugs. Genes do not stay put: transgenes spread, increasing the risks of viral resistance. (Geneticists prove this with the potato.) Fields of genetically engineered crops may lead to wildlife starving, for birds and insects often depend on weeds. Furthermore, most arguments in favor take little account of the effect of time. Nature, left to her own devices with time, regenerates, renews and reproduces. Biotechnology *neuters* nature, making it inert without artificial inputs, making nature in effect non-renewable. In one particularly nasty recent example using "Terminator Technology" a plant's seeds will not germinate if planted a second time. (These spannered crops were to be targeted mostly at farmers from countries of the South.) Vandana Shiva movingly described to me the profound conceptual tragedy—and sacrilege—this entails to an Indian sensibility. The word for seed is *bija* in Sanskrit, from *bi* meaning "source of renewal" and *ja* "life." Put the two together, she says, and you have "that which arises from itself within itself." The seed is the kernel of life's continuance, the pip—time incarnate—contains past, present and future within it. Terminator technology, the deliberate insertion of death into the very heart of life, stops time dead.

Many scientists express disquiet that genetic engineering experimentation gives humankind the power to cause the future, but such experiments seem to come from an atemporal way of thinking. In real time, they are flawed by two things: their irreversibility and their unpredictability.

People have been experimenting with crops for centuries, but indigenous experiments take place within time, over years and years, with the slow proposals of genetic changes, like evolution itself, allowing the subtlest mind—the mind of all species—and the longest memory—the memory of all time—to make its genetic propositions, within the whole complicated web of relationships, one tiny strand at a time. Genetic engineering experiments take place in abstracted lab conditions *outside* time and uncontextualized.

And now we have patenting of seeds. A company, by altering one genetic component of a seed, can patent it and sell it. And never mind that it has taken centuries of work and the passing-on of knowledge for the people of the Andes, say, to develop some 3,800 varieties of potato. Corporations come, seeing a potato as a property without a past and patent said potato, in effect stealing those years of work and the knowledge of indigenous peoples without any recompense. Here Western knowledge fails to comprehend indigenous knowledge which is accumulated over time—slow-knowledge. Western knowledge is fast-knowledge, seen as single acts of "discovery."

In indigenous thought, some things—of which knowledge is one—cannot be possessed privately. "Common knowledge," as English says, in that humble but profound term, is a metaphoric commons, like common land, owned by everyone, free to all, but possessed by none. It is a resonant, free and innocent phrase, a common expression, common as mud, as unprotected as it is unself-conscious. It assumes an intellectual right to roam on common landscapes of common wisdom. But this knowledge is now being privatized through patent laws. As surely as fencing land and nailing up Private—Keep Out signs, the metaphysical commons of the human mind are being stolen for private profit in a second, and arguably far more drastic, age of enclosure, enclosing common knowledge for the financial benefits of a few.

In 1548, the Enclosure Commissioners defined "the classic wicked enclosure" as "where any man hath taken away and enclosed any other men's commons" . . . Between 1760 and 1844, nearly four thousand Enclosure Acts were passed. Common land, for the common—public—good was thenceforth fenced off for the private good. Henry Fawcett, in his *Manual of Political Economy* of 1863 (published in 1876), wrote: "In the case of almost all these enclosures the interests of the poor have been systematically neglected."

History repeats itself, curling like a spiral, like a double helix, like DNA; new age, same pattern. What the historical enclosure movement was to land, is happening now to time, in many, many ways, from the enclosure of festivals to the enclosure of wild time, but the contemporary biotechnology movement is one of the most direct. Genetic engineering is the enclosure of time. The heritage of common genetic material is "fenced" by patent laws and made into a sellable commodity. The unstructured—open—sense of the past is numbered, sequenced and mapped.

During the period of land enclosures, fences meant thefts, the stealing of the peasants' common land. ("Fence," that cunning word, means both "artificial boundary" and "handler of stolen goods.") To add insult to injury, if the peasants later entered the land once theirs, they would be accused of being trespassers. With genetic patenting, a similar pattern emerges. Multinational corporations dare to ask for "protection" from indigenous peoples through patents; for, like peasants of old "trespassing" on land once theirs, if indigenous peoples today reuse seed which at just one stage is modified and then fenced by patent, they stand accused of intellectual trespass.

Biotechnology companies, like the twelve-billion-dollar-a-year Monsanto corporation, claim that only through genetic engineering will the world's population be fed and this will happen by companies holding the interests of their shareholders as primary. Genetic engineering will work for the common good via the private good. We've been here before. Landowners argued that through enclosure alone could new agricultural methods feed the population, that enclosure would work for the common good, via the private good—and they could not have been more wrong. The landowners of the enclosures are the Monsanto shareholders of today.

A neem, by any other name, would be as sweet a tree and as sweetly useful. Is it a toothbrush, a contraceptive, a moth–chaser or a candle, is it soap or a pesticide? All the above. A wonderful thing is the neem tree. For centuries, the Indian farmer has ground its seeds and scattered them, to protect crops from insect pests. Then, at the day's end, the neem tree farmer has washed his hands with neem oil soap, brushed his teeth with a neem tree twig, lit a neem oil candle, and gone to bed with his spouse— using neem tree contraception to scatter seeds of another color.

Many of the properties of the neem tree were discovered by Indian farmers, but two companies have obtained patents for derivatives of neem developed in their laboratories and are now producing *their* neem pesticides. In 1993, one company was given the patent on all future genetically engineered cotton. No one took much notice in the affluent West. In India, a wildly explosive demonstration took place; half a million Indian farmers demonstrated against W.R. Grace and other seed companies.

Nature—*Natoure*™—is ours, say the seed companies who sell modified seeds to Indian planters and (even without Terminator technology) can legally forbid those planters to collect the seeds at the year's end and replant them. The farmers must *rent* the seed; it will never belong to them. Time, here, is castrated, not allowed to run to seed. Nature is ransomed in serial cereal rent. Ceres, goddess of natural time—now bountifully begetting, now wintry fallow—is up for hire, anytime prostitute to the seed company's seedy pimping. (When farmers buy Monsanto seeds, they must sign a contract which forces them to use only Monsanto herbicides and bans them from resowing seeds.) The days are gone when nature and time could wander off for a frolic and a fruitful fuck, spillaging seed packets across the world. Humankind has sealed them off, turning the world into something half lab and half superstore. Nature is artificially inseminated by man, and time is forced to empty out his seed packet on the stroke of the commanding hand of man's clock.

Human nature is, of course, as much part of nature as the elements, the snow leopard or the hummingbird, and any discussion of time and nature must include humanity. If anything could be said to contain human time and nature, it is genes, storing the past and intimating the future.

John Moore is U.S. Patent number 4,438,032, the world's first patented man and not very happy about it: it happened without his knowledge or consent. In 1976, he developed hairy-cell leukemia and during the treatment for this, his doctor took samples of Moore's white blood cells, cultured them into a cell line, applied for, and was given, a patent for the "invention" of said cell line, then sold the line to a biotechnology company for 1.7 million dollars.

Moore now is less. "How," he asks, "has life become a commodity? I believe that all genetic material extracted from human beings should belong to society as a whole, and not be patentable." He speaks in the language of the Commons-Enclosure debate, seeing himself as a twentieth-century "Common Man." To describe the moment when his doctor, during the course of treatment, patented Moore's cell line, he even uses a metaphor of himself as land—"I was *harvested*," he says.

It could happen to you. If you're a woman, it has, in effect, happened to you. Women's bodies change dramatically, late in pregnancy; the birth canal alters and the cervix softens to ease the labor of childbirth. It happens due to a hormone called relaxin: in every woman's DNA is the genetic recipe for it. But this human gene doesn't belong to you or me or your mum, it belongs to Genentech, an American biotechnology company.

In 1989, the Hagahai peoples of Papua New Guinea were visited by medical anthropologists who took blood samples from them and patented the cell line of one Hagahai man. The patent (number 5,397,696) was granted on March 14, 1995, the first time that a patent had been given on cells taken from an indigenous person. The Hagahai people felt betrayed, robbed and violated.

The Human Genome Diversity Project (multi-government-funded) is establishing a gene bank of a hundred and twenty endangered peoples. Indigenous people ask one bitter question: Why don't they address the causes of our being endangered instead of spending millions to store us? For an answer, take the tomato, which used to grow in such variety—from the sweet-squirt of a cherry tomato, to that big fat splat of a beef tomato. Many years ago, Campbell Soup started to hoard tomato genes and is now in possession of a substantial portion of the world's tomato biodiversity. The "tragedy" of loss can be turned into the "prize" of scarcity in time as Christine von Weizsäcker points out in *The Life Industry*. In economic terms, it means "Buy now, sell later." For exactly the same reasons no company would mourn the loss of tomato biodiversity, provided they had banked their genes: so biotechnology companies will shed few tears for the extinction of tribes—the greater the scarcity, the more the gene banks will be worth.

The Institute of Genome Research is—reportedly—the world's largest human genetic data bank. In April 1994, the drug company SmithKline Beecham invested $115 million for an exclusive stake in this database. David King, a former geneticist who edits *Gen-Ethics News*, says: "In other words you have a corporation trying to monopolize control of a large part of the whole human genome—literally, the human heritage," and asks, "Should this become private property?"

Genes are the human heritage and, to many people, quintessentially a Commons, which should be unowned and unenclosable. Gene banks can be described as the enclosure of the past by private corporations, but genetics encloses, perhaps more ferociously, the *future*.

Genetic tests are not absolute prophecies of your future, but they show tendencies. Of course some genetic screening can be good—knowing a predisposition to certain diseases may mean a chance to prevent them. But after a few generations of genetic testing, some geneticists say, there could be not one common gene pool, but a divergence: those with "good" genes becoming ever more sharply separated from those with "bad;" a genetic aristocracy carrying synthetic genes, the "GenRich" as opposed to the genetic "Naturals," as Lee M. Silver, biology professor at Princeton University, puts it. Philosophy professor Philip Kitcher, in *The Lives to Come: The Genetic Revolution and Human Possibilities*, speaks passionately about the use of genetic foreknowledge, about the way it might divide society into genetic haves and have-nots. It is, he argues, the most important question facing society today. "The trouble is, no one seems to have noticed." The film *Gattaca* noticed this, but more tellingly, insurance companies, more sensitive than most to clouds on the social horizon, have heeded and already people shown to have a genetic predisposition to certain diseases are being refused health insurance and mortgages.

After the Enclosure Acts, the landowning classes became much more sharply demarcated from the landless peasants: in the same way, the "time-owning" classes of the genetically rich may be increasingly divided from the time-disowned classes of the genetically poor. There are benefits to come from genetics—the treatment of just one disease, cystic fibrosis, has been revolutionized by it—but every scientific advance takes place within a social context and these are times characterized by the strength of corporate interests. In such times, genetic benefits will—*must* by economic logic—be available only to those who can pay; to the land-owning classes but not to the landless peasant.

Time, in a metaphysical sense, can be mapped: maps of genes are, in effect, maps of time fencing time's infinity. Although genetic testing does not show the absolute, inevitable future, but only predispositions, it nevertheless casts long shadows over the idea of the infinite future. The infinite, in time, means mystery, chance or luck, your possibility, potential and hope. But if this infinity is mapped and fenced to finiteness by genetic "foreknowledge," there seems little point in pioneering, no reason for ambition, neither the reach nor the grasp. Your own time will have been mapped, fenced off and enclosed by genetic "prediction"—the future, which once rolled down to us like clouds of unknowing, won't roll. The very clouds will seem fore-counted, fore-known, fore-sold: the future a sale foreclosed. Echoing the religious conflict between free will and predestination, genetic "predestination" seems to overshadow free choice and tomorrow's shadows fall earlier than they have to.

Genes are said by some to show predispositions to, among other things, alcoholism, homosexuality, schizophrenia and manic depression. (With caution; no single gene has been found which influences depression, and the "discovery" of a gene which shows tendencies to homosexuality is debated.) But the important part is not perhaps the discovery but the desire to test people for such things. And the question is if society had already used such tests, who would have run the risk of permission refused?

"I am not a heavy drinker," Noel Coward would unsuccessfully plead. "I am not a heavy drinker. I can sometimes go for hours without a drop." Almost all artists, writers, musicians and comedians would be refused entry, genetically predisposed as they are to go on the brew, crack up and sleep with people they shouldn't. Baudelaire, Francis Bacon, Dylan Thomas, Omar Khayyam, Marguerite Duras, Virginia Woolf, Spike Milligan, John Cleese, Oscar Wilde, W.H. Auden and Gertrude Stein. All gone. You could have all five Spice Girls but neither the Beatles nor Beethoven. Five generations of Bachs, at a pinch, but barely one Brontë. O. Henry, Henry James, James Joyce, Joyce Cary— oh and Flann O'Brien, Brendan Behan and all other fellow geniuses of *genus* Guinness. Gone. Lifting the lid off a Liffey pub, the sun in its rising would light on none of these.

All these people of damaged inspiration, who make the whole circus of life worth the entry ticket—the lopsided ones, the margin-ivies, the myriad-minded and the orchidaceous— would be refused entry. Outcast by society and yet needed by society, the condition of the poet is both "the curse and the cure," according to Seamus Heaney, like the wounded bowman Philoctetes, outcast for his wound, but wanted for his bow. Genetic selection could mean neither wounds nor bows, neither difference nor metaphor, a society in uniform.

"Biotechnologies are, in essence, technologies for the breeding of uniformity," says Vandana Shiva. In nature, mono-varieties are dangerous because uniformity can leave crops vulnerable to disease and climate conditions: you can prove it with a potato. When the potato blight struck Ireland in 1845, it wiped out the entire crop because the potatoes were genetically uniform and had no resistance. Monoism destroyed the potato.

Monoism is the cultural mind-set of contemporary life; the mono, not *tri*umphant but *mono*umphant, succeeds from monohollywood to monomcdonald's, monopolistic companies and monopotatoes. Monoism threatens nature, by destroying biodiversity. And it also threatens future society by destroying the biodiversity of human nature through genetic selection.

Selection is a word which typifies much of this age: a spoiled consumer pushing the supermarket cart down the aisle of fetal selection; brown eyes or gay—*Tide* or *Shout*—blue eyes or straight—*Ivory Liquid* or *Palmolive*? At worst what the selecting procedures of genetic testing will produce is not humankind, but a shaving of it; for in the diddy bag of genes, humanity is the gorgeous cowrie shell, but is no less the dried nettle, the tumbleweed seed, the cracked button, the fluff in the seam and the hole in the corner.

Geneticists, far more than most lay people, are concerned by the social uses of genetics, for they, of all people, can never forget that its history is rooted in eugenics. Darwin heavily influenced the eugenicist Charles Davenport, who in the 1920s was the motivator behind the sterilization of some 25,000 Americans who were thought to be bearers of genes of "feeble-mindedness" or criminality. Of all eugenicists, it is Hitler whose legacy lingers longest, with 400,000 sterilizations and millions of deaths, in his name. The very title of Hitler's "Struggle"—*Mein Kampf*—is taken from Darwin's description of the "struggle" for existence. His eugenics silenced and sickened the world.

Eugenics is alive and well today. In 1994, the Chinese passed the "Maternal and Infant Health Care" law which sounds as benevolent as apple crumble and custard. Not so. It is a eugenic law requiring couples to undergo prenatal genetic testing and if a fetus is disabled or affected by genetic disease, abortion is mandatory.

For most of the world, though, eugenics needs nothing as unsubtle as law; the market is a far more efficient persuader than the state. The shampoo-voices of advertising will euphemistically coo: *You deserve a perfect child in a rinsed clean reconditioned world.* Surveys in the U.K. show that cosmetic genetic manipulation is becoming ever more acceptable; it comes politely, Avon-lady-like, tap tapping at your door and the advertising text has already been written, in a key of biting satire, by songwriter Theo Simon. In his song "Designer Kidz," the lyrics mimic the ad-to-come. (Hum fast.)

Goodbye fairy, cripple and yid,
Get a clone of your own designer kid.
Kids no doting mother ever had,
Created by "Selected Ad."
We're the market leaders in the race
To privatize the human face
We cracked the code of the chromosome
With a logo guaranteed on every bone
        . . . Diddleeeiiidum, iiiday, DNA.

And now we have cloning of sheep sheep.

Dolly the lamb hit the headlines early in 1997. Named after Dolly Parton, she was cloned from the udder cell of an adult sheep at the Roslin Institute outside Edinburgh. This cloning was a triumph over both nature and time; a man-managed "freezing" of time, resulting in Dolly's identical twin being born a generation after Dolly. Scientists engineered the udder cell of Dolly-the-elder to reverse the aging process and "remember" that it had the blueprint for the whole sheep from which to make Dolly-the-younger.

There was massive, international interest. In 1997, sheep can be cloned. When cometh the human clone? It will happen, say scientists, and soon. Germany, wise of its memories, was the most alarmed of all nations. The newspaper *Die Welt* wrote: "The cloning of human beings would fit precisely into Adolf Hitler's world view. And there is no doubt he would have used the technology intensively if it were available at that time. Thank God it wasn't." Atom-splitting, Nobel-prizewinning physicist Joseph Rotblat warned that such genetic engineering could result in "a means of mass destruction."

The cloning of Dolly was as important as the splitting of the atom, for the relationship between time and nature. Once, you couldn't split time and nature; you couldn't split the sunrise from the sun and you couldn't divide the dawning from the day. Now it has been done. It staggered the world and it staggered the sun itself. In natural time-frames, the sun rises once on the birth of identical twins. Now, the sun's course has been staggered, so it rose twice, first on the day when Dolly was born, second on the day when the clone was born. Metaphysical poet Marvell wrote: "We cannot make our sun stand still." *We* can. The day Dolly's clone was born, in a metaphysical sense, the sun was halted; the sun was stopped, that day.

Immediately after Dolly's cloning, the possibility of human cloning obsessed the public mind. (In the run-up to the millennium perhaps, to chime the digital replication of that zero-cloning year 2000 with human cloning seemed irresistible.) Many people initially confused cloning or genetic *replication* with resurrection or even immortality. It isn't. It would be the birth not of you but of your identical twin, at a different date. More accurate tut-tutters were quick to reprove the mistakes but, arguably, the confusion is itself the important thing, for society will probably continue to think a clone is a "replacement." How does that damage both a sense of personhood and time?

Cloning would foster one of modernity's cruelest tendencies, that of *replaceability*. If society confuses a genetic clone with an identical person, would murder henceforth be considered so absolutely immoral, if the dead could be supposedly "genetically replaced"? With replaceability always an option, would extinction be considered so bad?

Gregg Easterbrook in *A Moment on the Earth: The Coming Age of Environmental Optimism* writes: "The coming earth might be one in which extinctions are reversed . . . especially the recently extinct, for which the best tissue samples remain." General applause, perhaps. But that sentence has a ferocious shadow: *and therefore extinction doesn't matter.* Animals need not be protected, if they can be reassembled. Indigenous people can become extinct without even a sense of tragedy, because their genes have been banked. They are replaceable in re-re-re-productions.

"Eternity is in love with the productions of time," said Blake, and didn't think he'd ever need to add a footnote. (Fax arrives, from Heaven or Hell, Blake to his Editor-on-Earth: "Scan in *Irreplaceable* or *individual* or *unrepeatable* productions of time." Editor to Blake: "Line loses its panache. No scan do.") There is only one Blake. There is only one Dolly Parton. There is only two Dolly the sheep. Do you pluralize the verb *to be* for a clone? Are they one or is they two? If you're talking sheep, or sheeps, tongue strictly in cheek or cheeks, who could care less? If you're talking human beings, who could care more?

Irreplaceability is a time-concept of humanizing beauty. Every human value-system in art or in love is set to cherish the inimitable and to disregard the imitation; to protect the unique but not the replica; to notice the first edition but to ignore the photocopy.

And to love the specific person, the lass unparalleled, the irreplaceable you. *Irreplaceability* is an irreplaceably compassionate idea and its loss is a loss unparalleled.

To be truly full of life, each moment of time must be considered irreplaceable, unrepeatable. This sense of uniqueness, of the individual moment of time as irreplaceable as the individual human, is blunted by cloning; human emotions are given all their importance through time's uniqueness. If the sun is made to stand still; if time is "stopped" by repetition; then human passion has less furious vigor, human experience fewer contours of meaning.

Father Time and Mother Nature were once *in loco parentis* to the human race, the ultimate source, from whence—and from *when*—we all came. With genetic engineering, the parental role has been usurped by humanity so time and nature are not source but just *re*source. Furthermore, the fecund frolic necessary to procreation has been replaced by a sexless experiment. Post–Dolly, too, it is no longer axiomatic that mammals need a mother, a father and a broad quality of wink to procreate. Sex means that every generation tosses around new mixtures of genes, sex goes forth and multiplies the possible outcomes of birth and is vitally important to stop a species using up its genetic reserves.

Genetic engineering, on the other hand, prefers to circumvent sex altogether. And again, you can prove it with a potato. How fares the sex life of the potato? Steve Jones, geneticist, answers: "Every King Edward potato is identical to every other . . . This is convenient for the farmer and the grocer, which is why sex is not encouraged among potatoes." How fares the lay of the land, sexually speaking? "The rural landscape may become one in which asexual cows feed on engineered grass under the shade of clonal trees." If genetic engineering throws a spanner in the works of time and nature, sex throws a garter belt in the works of genetic engineers.

No scientist will promise that genetic engineering will work out well, though few absolutely predict all-out immediate nightmare; the majority agree on great caution. (Gregg Easterbrook speaking with characteristically blithe bubbles of flatulent optimism: "Genetic engineering may open the door to dystopia . . . it might result in habitats unlike any that would come into being through purely spontaneous forces . . . But as all habitats are guaranteed to change regardless of what *genus Homo* does, this alone should not deter," is very much in a minority.) But even if biotechnology doesn't produce dystopia, humanity has lost a quality of time—one which Virgil and Baudelaire knew—the *warmth* of the thing.

Dolly Mark II was born by a process of freezing cells: and it was the cell *clock* which was frozen. Genetic scientists plan to extract sperm from frozen mammoths in freezing Siberia and use them to fertilize Indian elephants. The mechanics of these processes require freezing but, more importantly, the very character of biotechnology inherently requires it. It is cold work, cold as clockwork, cold as the soul of commerce.

If mechanical clocks began the triumph over nature's time, genetic engineering perfects it, for this, like nothing else, freezes both nature and time, fixing an artificial clock in pride of place. "In the long run," says Easterbrook, promulgating genetic engineering, "there is no meaningful distinction between artificial and natural." Mind the man; freezing-cold mind at work on a cold cube of thought. His is a world created by artifice, with artifice and for artifice; death may have less dominion. Maybe. But life would be infinitely less warm to the touch.

There is such a thing as an assonance of temperature. Nothing humankind does, thinks or decides comes cold, every thought is an act of emotion; every decision a warm act of love or anger. Humans are warm-blooded creatures and we need warmth to survive, the warmth of the sunrise, the warmth of the sunset, the warmth of sex itself. Cometh the hour, cometh the man. In an inherently engineered hour, mankind can't cometh, can't live; in contrast to the warmth of the Golden Age, what we are engineering is an age of ice, wherein we freeze.

Watch your husband. If you are lucky—should he die in your presence—scoop him up and shovel him into the cellar; give him the kiss of half-life to keep his circulation going; stick a needle full of blood anti-coagulant into him; then cool his body with ice packs which you should have prepared earlier. Wrap him in blankets and freeze him with dry ice to minus 150 degrees Fahrenheit. Now move him into a capsule of liquid nitrogen which will freeze him to minus 320 degrees but—and this is the hard part—don't drop him, for should he accidentally slip to the floor (cold cellar tiles, frozen spouse) he will shatter like a ton of glass—a million icy shards of husband everywhere—terrible mess.

Then, rest in hope that cryonics, or "cryopreservation"—using extreme cold to preserve "living" tissue—will keep him intact, until future medical skills can de-ice him and reverse the cause of his death, whether the dicky ticker, the organ moribund, the circling C of cancer cells, curling like a finger, beckoning. (You say he died of old age? Nonsense. It is not legal to die of old age, no death certificate can say that; you *must* die in a clinical category.)

Of course, people have always wanted life after death, in heaven or in books, through fame or through their children. Of course, the insignia of dying, from death masks to funeral inscriptions all say *Hic Eram* (I was here). Of course, individuals have often wanted self-expression after death. One London art critic wants his remains mixed with bread crumbs and scattered outside the National Gallery where pigeons may eat and excrete them in an action painting which would be his final pooh-poohing of such art. A pub landlord wants his ashes put into an egg-timer so that silently and evermore he can be seen to call closing-time: "Time at the bar, now, please." There is, nonetheless, something about the new death industries—an almost cartoonish quality. Cryonics is also called "suspended animation," and, rest assured, Walt Disney, king of cartoon, is frozen in California, awaiting his full feature re-animation.

Trans Time is an organization dedicated to seeking ways of extending the natural lifespan. Based in San Leandro near San Francisco, they charge a variable insurance policy, a fee of, say, $300 a year for a healthy forty-year-old man and, for this, these undertakers of ice undertake to freeze you in perpetuity. Art Quaise, their president, says he believes that "medical advances in future will be able to cure all diseases, and be able to halt and reverse age." For, as he says, "Isn't it better to be young and virile than decrepit and aging?" Not for him the ineluctable democracy—even humility—of death. Cryonics is "about realizing that you are the most important person on earth. Nothing else counts unless I'm around to have it count," he says.

Marina Benjamin, in *Living at the End of the World*, writes of visiting the Alcor Life Extension Foundation (in Scottsdale, Arizona), the largest cryonics organization in the world, and talking to Steve Bridge, President of Alcor in 1996. He worries his eventual defrosting may leave him with a body but without a personality. He has taken precautions. "I've asked all my friends to save my letters and I've stored a huge amount of stuff on computer disks over the years. I have access to a lot of who I am, so if I come back minus my memory I can regain part of it by reading about who I was." Benjamin wryly remarks of this and other conversations with cryonauts: "It may well be that tomorrow's immortals will be as disappointing as Swift's everlasting Struldbruggs, who Gulliver expected to be rich, wise, virtuous and just, only to find that 'they were not only opinionative, peevish, covetous, morose, vain, talkative, but incapable of all natural friendship and dead to all natural affection.' "

Cryonics is only one option for those who would live forever. David Pizer, a real estate investor in Phoenix, Arizona, had invested large sums in cryonics, so that he could be frozen and reheated later, pizza-like. Now, though, following Dolly the sheep, he wants to be cloned. Like Quaise, he stresses above all the survival of the *self:* "I want to be able to live forever, in some form, some place. I want either myself or an exact duplicate copy— and I mean *exact* duplicate—of myself to exist." (Richard Dawkins, appropriately author of the "selfish" gene theory, has also—though not very seriously—expressed a wish to be cloned.) In Seattle, a company called Immortal Genes offers eternity in a paperweight. For $50, it will preserve your DNA in a little box for the next ten thousand years from which you can be cloned at some later date. If things go wrong there is also a ten-thousand-year money-back guarantee.

All these phenomena might appeal only to a tiny minority, but they are telling features of modernity's way of death, not typical, exactly, but symptomatic. They reveal, by exaggerating, these current themes: a steep concern with the individual life, a hatred of aging and a reliance on professional expertise and medicalization. Moreover, they illustrate a refusal to join the cyclical aspect of death, so instead of being buried in earth, the body reabsorbed into a larger natural system, flesh to clay, they opt for an offground—and linear—event, the individual freezing herself or himself out of society, on a solo mission to escape the compostheap communality of death by means of a fragmented individual "immortality." (This is symptomatic of the way modernity treats time, too; denying both its cyclic character and its earthy relationship to nature.)

Death, the first Democrat, the great grave Digger and Leveller, is universal, but there are no universals in the ways different societies treat death. Some celebrate it splendidly, some shudder: some won't remember, some won't forget.

The Dogon, of Mali in West Africa, have magnificent communal rituals to celebrate death, massive masked dances with wild impersonations of animals, trees and spirits. Neighboring tribes are satirized, as are white people. Here is someone in a "whiteman's-mask" who does not dance but walks stiffly, an "administrator," carrying paper and pencil and writing demands for money. There is an "anthropologist" who sits on a chair waving a notebook and asking the daftest questions the Dogon can possibly think of. Here the "tourist" with a camera, pushing everyone out of the way. These dances are pure vitality, in praise of the entire human comedy; a lifeful mask over the face of death.

The Mexican Day of the Dead, with its feasts for the spirits, skeleton puppets, plasticene trees of life, skulls made of sugar and skeletons playing dominoes, is equally an uproarious celebration of life. You mustn't cry, they say, on the Day of the Dead because it makes the road slippery and treacherous for the returning soul. Italians have their Day of the Dead. The dead are remembered in November, the cemeteries and tombs cleaned and food offered to the ancestors.

The Achuar, headhunters of Ecuador, by contrast, hate and fear their dead. They call death "a spear of twilight" and the dead are given no tombstone, honored with no memory. They are ostracized and expelled from the mind as quickly as possible; the living, they believe, cannot be fully alive unless the dead are truly dead. "Dwelling In Tombs Is Strictly Prohibited," says a sign on caves, in Golconda, Hyderabad, in India. It is a literal injunction many societies across the world, such as the Achuar, also understand for its metaphoric message.

But not all. In a Torajan house in Sulawesi, you may find that a motionless bundle of old clothes in the corner of the room is actually a dead grandparent. Food and drink may be placed on the body, or it could even be used as a convenient shelf, in one instance for a collection of tape cassettes.

In sharp contrast, British Gypsies, rigorously separating the living and the dead, want to continue their tradition of torching the caravans of their dead elders, though the British authorities (who also want to continue their own tradition of torching the culture of Gypsies) ban the practice.

The Pirá-paraná of northwestern Amazonia also sharply divide death and life; the dead are wrapped in a hammock and placed in a canoe. A woman is buried with her basket containing paint, a mirror and other personal things; a man with his ritual dance ornaments, feather headdresses and monkey-fur tassels; any other property of the dead is burned. The shaman demarcates a line between the living and the dead, by performing rituals with smoke and burning beeswax and snuff. Death over, life begins again.

Aldo Massola, in *The Aborigines of South Eastern Australia As They Were,* describes how, after death, the spirit went to the Land Beyond the Sky, reached by being blown on the wind, by jumping from a high rock, climbing a tall tree or a wallaby-sinew rope or by clambering up the rays of the setting sun. After a time there, the spirit returned to its totemic center on earth and, entering the body of a woman, would be reborn as a child. Believing in the indestructibility of the spirit, "alternate life and death was like day and night. Aborigines were not afraid to die." In Australia, just as the Dreamtime Ancestors did not die but returned into the land, so "each Aboriginal person intends . . . to sing himself back into the land," writes David Abram in *The Spell of the Sensuous.* A traditional Pintupi man will return to the place where he was conceived, his Songline, to die, so that his vital spirit will rejoin the Dreaming there.

In a gem of an image, the brief glory of life—this bright and peopled, tapestried, firelit banquet—was described by the Venerable Bede:

> As if when on a winter's night you sit feasting with your ealdormen and thegns—a single sparrow should fly swiftly into the hall, and coming in at one door, instantly fly out through another . . . from winter going into winter again, it is lost to your eyes. Somewhat like this appears the life of man.

These snapshots of death are distinctive, but they have common themes; the acceptance of death as part of nature's cycles; a sense of a community of the living which is stronger than the individual death, and a love of life which is expressed at once in a hatred of death *and* in its lively celebration. Against such a backdrop, contemporary Western dying is unusual. The subject is blanked out, a forbidden, freakishly unfamiliar topic. When British TV recently decided to show someone dying on screen, it was described as the "last taboo." Everyday life is wiped clean of death. It is displaced from the ordinary: fifty-four percent of deaths in Britain take place in hospitals and most people die alone, without family or friends, without any sense of the community of the living (a shift which came about between 1930 and 1950).

Anthropologist Nigel Barley, in an often hilarious book on death, *Dancing on the Grave*, describes a "post-mortem video," designed to be watched by the relatives after such a death: a frail old woman, on the point of dying, sitting for the camera in a pink bedjacket, "mouthed the usual orthodoxies celebrating the togetherness of family life and the values of contemporary America while alone in a solitary room in a hospital." At the end of it, she looks up at the camera muttering, "Is that enough? Is that what they want? Ah, what the hell. You're all full of crap."

The displacement of death is not only a social removal from the family into the isolation of professional care, but also a displacement from the earth, the raw intimacy that exists between the dead human body and the natural world. The earth is seen as "mother" to so many cultures and, in widespread traditions, a newborn baby was put on the soil of its mother-earth. Similarly, there are many traditions of laying the dead on the earth, completing a circle. The Chinese traditional funeral said: "Let the flesh and the bones return once more to the Earth." The Romans expressed the all-but-universal desire of the dying to return to their native soil—"native," of course, comes from *natus*—born. In dying, you return to the very earth from which you were born, which gives such depth to the inscriptions on Roman tombs: *hic natus hic situs est*. (Here he was born, here he is laid.) This instinct, like that which drives salmon to return where they were born, so often guides the old back to die at their birthplace—even Christianity's influence cannot altogether silence its earth-bound and therefore quintessentially pagan, resonance.

"Earth to earth, dust to dust:" the words of Christian burial pay funereal lip-service to the profound—organic—link between the dead body and the soil, but that's about it. Christianity, removed from pagan mud, has an afterlife which is clean of composty soil, heaven is an "unearthly" abstraction in a vinyl everafter. In a creepy advertisement on Christian television, a camera shot of a young—and exceptionally clean—man in a dinner-jacket was captioned "All dressed up for the most important day of your life." His wedding? The birth of his child? you wonder. The camera pans back and you see he is in his coffin. This seems all too characteristic of Christianity, which considers death more important than birth and thinks the afterlife is more precious than this muddy earthly life. (Cleanliness being next to godliness only because muddiness is next to paganism.) This attitude was epitomized by Teresa of Avila who openly desired death because she longed for eternal life: "Oh death, oh death, I do not know why you are so feared, since it is you who contain life!" After her death, her devotees declared that her body did not decay and was "incorruptible." For a linear religion which has long hated nature's fluxy ways, especially women's fluxes, Teresa was an icon of perfected flesh. Her very body was thought to have shunned nature's cyclicity below for the incorruptible heavens above. She would be marble not mold. The same ideas prevail today: "Above-Ground. The Clean Burial. Not Underground with Earth's Disturbing Elements" runs a full-page newspaper advertisement for a mausoleum director in contemporary America.

Hegel suggested that death rites were constituted of two opposing tendencies; the "fusion" and the "recoil" with respect to the earth, the embrace *and* the rejection of it. The instinct of recoil can be seen in the frequency of deaths in hospital, in the whole cryonics movement and in society's general denial of death. Walt Whitman in *Leaves of Grass* expressed the instinct towards fusion:

I bequeath myself to the dirt to grow from the grass I love
If you want me again, look for me under your bootsoles.

Montaigne said he wanted to die suddenly, digging cabbages. My mother wants to be left after death on the compost heap in the garden she has loved for some twenty years. Germaine Greer was recently asked how she would like to be remembered. "Compost," she likewise replied. "I'd want people to say she made good compost."

The recoil is today vastly more in evidence than the fusion. But arguably it is only when the fusion is accepted that death loses its terror. For myself, I know I would rage and rage against death unless I could lie down on the soil of the earth itself. Then I wouldn't mind so much.

This instinct for fusion is both place-related—the bootsoles, cabbage-patch and compost-heap—and *time*-related—to fuse with the seasonal time of the earth. Native American Daniel Zapata says: "We accept death as a part of life. Our lives, our seasons, they rise, they fall, they rise, they fall. We are like the seasons. We come and go." The Arakmbut of Peru remember their dead in one month, the month of the rivers' rising, so human death is tied to nature. The ancient Babylonian *Epic of Gilgamesh* describes the quest for human immortality: in one fabulous, filmic image, Gilgamesh dives to the bottom of the cosmic sea to pluck the "watercress of immortality" but what he actually discovers is the story of nature's time and seasons.

British undertakers have found that the pictures on their walls most popular with mourners are those of the turning seasons, particularly autumn, the paintings of Constable and his trusty, rusty oaks, which acknowledge a dying while placing it within the cycle of nature. (This ties in with the Celtic belief that certain times of the year—particularly the late autumn which we know as Halloween—were "thin" times when humans were closer to the spirit world.) For the easiest way to accept death's human finality is to contextualize it in nature's time. When my last grandparent died, one November, that month of dying, I was frantic to find autumn flowers, of rust and leaf-fall colors for her coffin, and thus to bring her death symbolically into the living pattern of nature, its rise and fall, rather than being a death in exile, a plastic moment in a hospital time.

Cryonics represents a spectacular rejection of any such earthy recycling: those frozen bodies (sometimes referred to as "suspendees") are shot high-tech on a linear trajectory into the future, outside time and alone. An ice-arrow isolation, compared to the warmer composty commons, the companionship of a death in earth. Cryonics is heir to Christianity's linear-time, both opposed to any circularity, whether the gardener's pragmatic compost-recycling, or a spiritual reincarnation. Satish Kumar remarks how this "contrasts with the traditional wisdom of India where the life   327 force is seen as moving in circles, where human beings after death can be reborn as animals, birds or trees. Life does not end in death." (Besides, he adds, wryly, if you're coming back to this world, you're a little more careful how you leave it.) Sure-yani Poroso, of the Leco, says: "We have no word for death in our language. Time is not inert. Life returns, after death. The person returns, it is all fertilizing. My grandfather lives in different plants and fruits and from this thought comes a profound respect."

John Muir writes of "the sympathy, the friendly union, of life and death so apparent in Nature . . . the beautiful blendings and communions of death and life, their joyous inseparable unity, as taught in woods and meadows, plains and mountains and streams of our blessed star."

Western society ignores and hates death, treating the old nudging near to it with dislike and scorn; in contrast to the ancestor worship of many cultures. In Korea, the gatherings for ancestor worship at fixed times of the year were by tradition an important part of the cultural calendar. Genealogies circulated in Korea from 1600 or so, and were, in the 1930s, still the most frequently printed publications. In the West, the miserable moribund are "the elderly," not, for instance, "the elders." Snake-oil salesmen touting eternal youth become superstars: people not only buy Deepak Chopra's *Ageless Body, Timeless Mind*, they even read it. Amongst the Beaver people of the eastern foothills of the Rocky Mountains, the phrase "Old-timer" is used to suggest that the elders are sources of wisdom. Cezar, an Arakmbut man in Peru, comments: "Our old people are like a library, or like the roots for us." In the West, with its cultural obsession with technology and the new, the elders are often considered less wise than a twenty-year-old. It seems slightly perverse that we are devaluing old age at the point when people, in the West at any rate, are living longer and longer. State of the art death-technologists offer an exaggerated register of these wider social attitudes. Art Quaise speaks with contempt of becoming "decrepit and aging," and the frozen David Pizers, all of cloned him, look forward to a future which can (micro)wave a wand to turn age to youth. They, like society, are blind to the benefits of age, its wisdom, long perspectives and its capacity for exuberant, naughty, life-enhancing shamelessness—the very old can be far better than rebellious teenagers at breaking social rules. Other cultures are kinder. Some have gerontocracies—the rule of elders. In India, *bura* means "old" and also "wise" and "powerful."

Modern death is not for amateurs; it is a business, a matter for paid professionals, for the experts. (Cryonics *et al* exaggerates those characteristics, being both highly technical and very expensive.) There are bereavement counselors and professional undertakers with a selection of priced goods. Grief management is on sale. Death, from coffins to hearses, is packaged. This, be it said, has something of a history. In the sixteenth century, the parish clergy in Madrid took on many more masses than they could hope to actually say. The situation was worsened by economic inflation. As the cost of living increased, so, in effect, did the cost of dying. The masses paid for at lower prices could not support the clergy some years later. Masses went unsaid, and the rich grew concerned to buy their way out of purgatory. Consequently, by 1596, those who could do so began paying a price for masses higher than that required: "creating, as it were, a black market in masses," as Carlos Eire writes in *From Madrid to Purgatory.* A generation ago, in Mexico, says Ivan Illich in *Tools for Conviviality*, the only professionals involved in death were the gravedigger and the priest. "At first, undertakers had difficulty finding clients because even in large cities people still knew how to bury their dead." Now, "legislation is being passed to make the mortician's ministrations compulsory." Today, the technologization of death has meant that in the U.S., the bereaved can go to the funeral in cyberspace, can see the coffin, send flowers and talk about the dead on the Internet.

In terms of medicalization, the omega of death is only matched by the alpha of birth. And there is protest. Like the Natural Birth Movement there is in Europe and the U.S. now a Natural Death Movement. The U.K. has a Natural Death Center with a "Manifesto for the dying," which instead of letting death be "packaged" in standard supermarket models, aims to let the dying choose their fall, to find the death which is theirs. If modernity's way of death denies nature, the Natural Death Movement re-emphasizes nature: trees to mark a grave instead of a tombstone and symbolic rituals which tie the dying into seasonal cycles. They also campaign to de-medicalize the process and to promote "midwives for the dying"—skilled companions to those tiptoeing out.

An English Day of the Dead has been suggested, to relocate death within the public and cultural calendar. Bereavement can be shockingly lonely, because modern society denies death so fiercely, seeing it as a personal matter, private to the point of isolation; grievers must grieve alone, the dying must die alone and in a hospital death you may not stay with the body no matter what instincts are keening for you to do so. (When my grandmother died, we were not allowed to stay with her. I felt I could, somehow, warm her with my body, with the warm community of the living at the instant of her death. Usher away such *communitas*.) It was not ever thus. Death was communal once; until the seventeenth century, the rooms of the dying and the dead were often crowded with people, including children. But the seeds of today's isolate customs of dying were sown early. From the eleventh century, according to historian Philippe Ariès in *Western Attitudes Toward Death*, "a formerly unknown relationship developed between the death of each individual and his awareness of being an individual." From the thirteenth century, tombs were increasingly individualized, with inscriptions and effigies and, from the fourteenth century, the death mask.

Other societies go collective. Maori writer, Witi Ihimaera describes the Maori way of death in *Tangi*. The Maori family, he explains, means "not only family living but family dead," and the past ancestors are "ahead" of you. People stay with the dead body for several days; a death is an event for the whole community. The whole ritual leads to a stronger sense of the *continuation* of time—the community's time—not the *end* of time in the individual's death.

In Madagascar, the Merina have a death ritual which involves the grinding up and mixing of dead bodies; so that, in effect, the individual defeats death's finality by being reformed collectively. In Cameroon, in Dowayo ritual, skulls are jumbled up in jars; the individual has died, but becomes part of the collective ancestorhood. (That point, says Nigel Barley, when the dead individual is dead *as* an individual, is when survivors can champion their own individuality. Widows, for instance, sing: "Hitherto we have all lived together. Now I shall fart in my hut and you shall fart in yours.")

There is a parallel with myths, which is not altogether surprising, for myth-making is, in great measure, a way to comprehend death. In myths, the individual story ends, but the collective life continues; as in the Maori, Merina or Dowayo ways of dying, while the individual life ends in death, the life of the community goes on.

The West attempts, almost operatically, to prolong the specifically *individual* life. It is the individual—the self—which Art Quaise and David Pizer are so zealous to preserve. (And, you have to ask, how many living bodies in wider society could be cared for by all the money being spent on preserving a few over-iced, over-priced dead bodies?) There is a hysterical opposition to euthanasia and to its practitioners; Jack Kevorkian, a Michigan doctor specializing in euthanasia, has been hounded through the American legal system and subjected to endless hate-mail. Such hostility leads some to promote an alternative to medical euthanasia; a voluntary fasting-to-death as by Native American tradition.

The violent opposition to euthanasia and the fantasmagoric cryonics-*cum*-cloning ideas show an incredible wish to prolong individual life; it is as if the individual's desire to live forever has become overwhelming at precisely the moment when it looks possible that we as a whole human community, as a collective species, may not. (Not just for fun did Levi-Strauss define the study of mankind as *Entropologie*.)

Death has an inexorable generosity: each generation takes the stage but eventually must bow out to give room to the next, rotting and recyling. But, as a result of the Western way of death—combining Christianity's hatred of recycling life and the impossible demands of individuals to live forever, or at least to be represented in mausoleums forever—something rather anally retentive happens with death. In the overcrowded churchyards of Britain, there is little space left for those dying today, because the dead of generations ago still monopolize the burial sites, straddling beyond their time with all the buildings of death, whether marble or stone, mausoleum or crypt or headstone. There is too much Edifice in this death; the stone orchards, with their marble branches. Planting a tree for the dead seems kinder; for all trees are trees of life.

In another sense, modernity desires to represent itself forever—in the storing of endless photographic archives. Looking at a personal photo album, supposedly of youth and happy memories, can actually make people severely unhappy, functioning as it does as a twentieth-century *memento mori*.

Yet another way modernity is unusual regarding death is that in its full process there is something akin to the erotic—the seeding recreation and conception. One beautiful rendition of this subtle sexuality is in Bill Viola's installation *Heaven and Earth* (1992). One screen shows the face of a dying woman, the other a newborn child; they occur simultaneously, they seem to immerse themselves in each other, as the two screens literally and conceptually reflect each other, life in death and death in life. The work is dedicated to his mother who died in 1991 and—death conceiving life—to his second son born nine months later. Between the sixteenth and eighteenth century came the beginnings of the eroticization of death; Thanatos and Eros, Death and Sex. The climax of Romeo and Juliet's great love is set in a tomb, Bernini's Teresa of Avila is pure sex and pure death, her orgasmic gasp that of the death agony. (An orgasm was once widely known as the little death.)

English sociologist Geoffrey Gorer, in an article titled *The Pornography of Death* (1955), describes how this sexual element shadows us this century; that mourning, because it must be private, solitary and somehow shameful, is like a morbid sort of masturbation. As death today is the principal forbidden and rejected subject, so, inevitably, hard on the heels of the forbidding comes the sin itself; Thanatos and Eros in the gruesome coupling of TV sex'n'violence, and in what is probably still the most horrific example, in the pornographic film *O* a woman was killed on camera for viewers to masturbate over her murder.

When death is used for pornographic violence, the deepest sense of its sexuality is traduced—the regenerative potency of time which needs a death to reseed itself. Time is charged with rebirth only by death. There are notorious stories of copulation at funerals, ancient and modern (more affairs start at funerals than you might like to think) and prostitutes in northwest Europe used to solicit in graveyards, because there is a resurgence of sexual energy even or especially at the point of death.

Time does not die, time is regenerated at and because of death; spring wouldn't spring if autumn didn't fall. Days of the

dead by common consensus fall at the end of October and November's beginning—Halloween and the Mexican and Italian
Days of the Dead—the year's polar opposite to May and May Day.
If there were no November, there could be no May; no month of
sex and couplings without one of partings and deaths. That is
the wisdom of death, so bleak, so sadly fair, so ruthlessly generous,
with its sincere severity and its sad, sad austere beauty.

13 · WILD TIME

The gods confound the man who first found out
How to distinguish hours—confound him, too,
Who in this place set up a sun-dial,
To cut and hack my days so wretchedly
Into small pieces.

—PLAUTUS

Life loves the liver of it.

—MAYA ANGELOU

The Alaska State Museum in Juneau is a small, squat and shrewd affair. Downstairs, it has an exhibition of indigenous people's lives and livelihoods, coupled with—when I visited—a touring exhibition of black-and-white land-portraits taken by photographer Bradford Washburn. For eons, the Tlingits, Tahltans and other peoples have lived in this wilderness Washburn portrays, their lifestyle unchanged for generations. Some of these pictures were taken in the 1930s, some in the 1970s and it is impossible to tell which was taken when, not a hint that "time" ever came here; no history seems to have blown across these valleys and rivers. These are images of an unmarked land without boundaries or limits, where time is unmarked and limitless.

Then, upstairs, Clock Horror! In the 1700s, the Russians came, fighting the Tlingits for territory. In 1867, Russia sells Alaska to the U.S. In the 1880s, gold is discovered in the Gastineau Channel. In 1896 there's gold at the Klondike. The exhibition has moved from limitless-time to Time Limited, with the arrival of white people, particularly in the Alaskan gold rushes. In a mock-up of an early bank, the First National Bank, there is a big clock and a calendar on the wall. (Saturday 21 April.) The writing is on the wall; Western Time has come to the wilderness (counting on finding money, its alter ego).

Time in a gold "rush" is critical: the speed with which a miner prospects, stakes a claim and registers it, makes the difference between a millionaire and a broken man. Significantly, the bank is the *First* National Bank, for this place rapidly became the epitome of the Western race to riches; the first shall be first in wealth, the last lost entirely.

The limitless time of the wilderness is now spattered with temporal exclamation marks; *Gold has been discovered! Fifteen day clean up!* shout the notices. "GOLD RUSH!" cried the newspapers, to excite the stampeders. Hurry, hurry, hurry! Boom and bust! Boom and bust! 1896, 1898, these golden dates fleck and glitter in a goldpanner's pan in the deep dark river of time; these years—weighed to the month, the day and the hour—brittle and dazzling, mark a spot and stake a claim of date and time, plot the previously unplotted place and mark the previously unmarked moment, for these "invaders of Nature," as a contemporary newspaper called the miners, imported two new ideas; the desire to limit, enclose and own the land and a sense of the limited nature of time, of a clock to be raced and beaten.

The museum's Downstairs time and Upstairs time are sharply opposed. If Plato had gone to Juneau, he might have seen a manifestation of his twin ideas of Being (Downstairs) and Becoming (Upstairs). It is easy to distinguish them by calling the Western chronism Time and to contrast it with the Timelessness of the indigenous land, the wilderness, but this would be inaccurate. Indigenous peoples have a far acuter sense of nature's time than industrialized Westerners. There is clocklessness, for sure, but no such thing as "timelessness." The wilderness is full of the phases of *time itself,* its qualities, its changes, its bear births, eagle marriages and salmon deaths. It isn't Western time, but a wild and untamed time—and perhaps just as wilderness describes this land, so "wildertime" or "wild time" could describe time here.

I left Juneau that afternoon to go into the Taku, one of the world's great wildernesses; four and a half million acres of wild land and four and a half million years of wild time. We rafted down the Taku River watershed—three rivers, the Sheslay, Inklin and Taku rivers, running 280 kilometers through British Columbia and Alaska—and I am writing this now, seven days later, from a valleyside high above the Inklin. On the opposite valley, a half-mile swathe of spruce has been felled, scattered like matchsticks by a brief and petulant tornado. I sit writing this on a dead-straight fallen aspen tree, silver, softened by ice and searing sun, softened so its bark is a silver pelt, polished, smooth as silence.

In the raft-days behind me were volcanic peaks echoing with thunder; grassy pastures zipping with cicadas; pastures where roses, sage, alpine strawberries and juniper, with foggy-purple berries and a smell of sweet extravagance, bloomed; and the river ran through box-canyons of gargantuan Homeric water which hurls rafts against cliffs and sucks them round whirlpools. In the raft-days ahead will come the mystery of a massive limestone mountain with underground streams; a sixteen-hundred-foot waterfall that runs so fast and falls so slow; and finally the ancient glaciers, place of blue ice and—inexplicably— *ladybugs.* (As red and as evanescent as ladybugs are poppies, and they too can bloom along the ice foot of glaciers.) Here, across the valley, is a vast, curving, unimpeded parabola of a world, in uninterrupted flow and as, across the summer, across the world, children blow dandelion clocks and seeds of fluff fall from unfluffy skies, so here, for an unremarked mile and more, one piece of seed fluff floats in a line of its own calling, choosing an arc of its own describing, like a dandelion clock choosing the hour. (The hour of now, the only hour the dandelion deigns to name.)

What is wilderness? Nature without Audience. That which describes itself, but which is unnamed by man. The British poet Robert Service wrote of the neighboring Yukon province:

There's a land where the mountains are nameless,
And the rivers all run god knows where.

The act of naming is an act of taming and in the Taku some rivers and mountains are named, some not. Some have Western names, some Tlingit, and the difference between them is telling. The Western names include Mount Lester Jones, or Wright Peaks. Tlingit names include The Sleeping Giants, and Taku itself is the onomatopoeic representation of "where the swans or geese touch down to land." Even in naming, these are two ways of thinking about time—one the Upstairs up-tempo Western naming which characteristically ties the object into individual biography and actual history—the tamed past. Tlingit naming, Downstairs, down-tempo time, ties the mountain or river into mythic biography and an abstract history—an untamed past which is also present: the sleeping giants still sleep, and the Taku is where swans and geese still touch down.

As an unnamed place is an untamed place, so an unnamed time is a wild time. The Taku is a wild river and, as rivers by almost universal analogy represent time, a *wild* river is a perfect setting for *wild* time. Things happen here, it is far from uneventful, but the trees torn up and tossed to the river bank are unregistered and unrecorded. A whole forest falls without anyone ascribing a date. Entire cliffs fold their stone robes and slide into rivers without anyone clocking the time it happened.

For years I have wondered what it feels like to be in a wilderness. Wilderness is a ferocious intoxication which sweeps over your senses with rinsing vitality, leaving you stripped to the vivid, your senses rubbed until they shine. It is an untouched place which touches you deeply and its aftermath—when landscape becomes innerscape—leaves you elated, awed and changed utterly. Forget the lullaby balm of nature tame as a well-fed lawn, here nature has a lean and violent *waking* grandeur which will not let you sleep. Cultural synonyms for wilderness in dictionaries and thesauruses list: waste, space, useless, barren, virgin land and seclusion. These are perniciously inaccurate. It is an aphrodisiac; it is a place of furious fecundity; not one of waste empty space, but of such ripe fullness that not four and a half million acres will contain it; not a place of seclusion but of rough engagement; not virginal, but erupting with the unenclosable passion at the volcanic heart of life.

But perhaps there is a reason for the mismatch between my perception and that of the editors of the *Oxford English Dictionary* and *Roget's Thesaurus*. Theirs is a once-correct, but now antique, conception, reporting back from a disappeared world. For, in the past, "wilderness" was something huge and vast which surrounded humanity. Wilderness was the Condition of the world within which mankind lived in perplexed pockets, plotting our little patches of garden plots hard by the great forests of wilderness. But then the human race gigantized in development, exploded in population across the world so that, past a critical point, wilderness and mankind changed places. It is now something *we* surround; there are pitiful pockets of wilderness dotted across the world; wilderness is now the Exception, and mankind the Condition of landscape. (This critical point for land, when wilderness ceased to be the Condition and became the Exception, happened, according to wilderness poet Gary Snyder, some ninety years ago; exactly when "time," as this book has argued, was undergoing such profound transformations.)

To me, this is a model for our relationship with time: for, once, humans were surrounded by wild time and the stretch of time was everlasting, undefined, unenclosed, unnamed, unchartered, a mystery lasting longer in all directions than even the longest evening which never ends here in the land of the midnight light, and into this eternity mankind was dotted, pitiful with our perplexed pocket watches and our brief lives, plotting our little watches of hours hard by the great eternities of wild time. Then we began to chart time, to clock it, plot it, measure and mark it, buy and sell it. As wilderness and humanity changed places, so too have wild time and mankind now swapped positions. Past a critical moment of moment-measuring, Western society's peculiar time-marking has become standard, a norm; the Western clock the Condition of time and wild time the Exception.

This can be illustrated on a personal level; take the company of clocks and watches. At first, the clock can be a comforter, the regular tick-tock like a mother's heartbeat to the baby in the womb; the boxy security of a grandmother or grandfather clock is a soothing presence in a room; and the tucked-up time in a pocket watch offers the cosy, looked-after feeling of the best parental care. When my mother gave me her watch, it was charged with her talismanic protection. See also how old people need their clocks. (Jessie, an eighty-year-old in hospital, has three things by her bed; her slippers, her teeth and her clock.) Older people need that same parental comfort—age, after all, is a place of second childhood. The company of clocks which tame wild time is a security in the face of the vastest aloneness: pure time, wild time with its wind-in-your-hair freedom, is exhilarating to the young, but can be frightening to the old, for wild time's darkest side is death itself. There are no clocks in heaven—or in hell. The clock of Frederick the Great is said to have stopped when he died, as the song "My Grandfather's Clock" has it: "The clock. Stopped. Never to go a-gain, when the o-ld m-an d-ied." But did it? Maybe it was the other way round. Maybe the clock stopped first and, suddenly, without his companion's genial tock, the old man lost the steadying beat marking out a regular rhythm for him, lost his step, stumbled over a gap in time and his inner ticker stopped.

So much, then, for the comfort of clocks and the security of watches, because what starts as a consoler can become a bully. Thinking to use a watch to control wild time, the watch-wearer may find that it controls them—the clock ordering the human, the bossy watch coercing its wearer, the commanding clock-time which is such a feature of modernity. The clock has become the Condition of tamed time which is overruling the Exceptional wild time.

Wild land and wild time were both charted, logged and discovered with the aid of theodolites, chronometers and telescopes; inventions all made, incidentally, in the same period of history; objects of finding in an unfoundland, inventions designed to find and log an unfoundtime. "Logging" is a heavy word today, its meanings are many; to log is "to find and make inventory" and "to fell forests." Once it possessed such resonance of security in a pitiless world (a log cabin and a log fire), but it has now, past our critical moment, picked up the overtones of pitilessness itself, as rainforests are logged and wildernesses brought to their knees. The Taku is threatened by this. A mining company, formerly called Redfern Resources now Redcorp Ventures Ltd., wishes to build a road across the wilderness to truck out copper ore, and logging companies are stampeding to negotiate the use of the road to log—in all senses—the Taku; both to find and make inventory, and to fell its forests; to find the land and to lose it at a stroke. Logging has become a threat to wilderness rather than a security from it. Likewise, roads and tracks, once paths of safety across a hostile world are, past a critical point of development, themselves a threat. As literal tracks, paths and roads shrink wildernesses, so similarly clocks make endless tracks across hours, shrinking wild time. Robert Marshall wrote passionately against proposals for roads in undeveloped areas in order to preserve a "certain precious value of the timeless, the mysterious, and the primordial . . . in a world overrun by split-second schedules . . ."

Would you look at a river and say it was running out? Rivers don't run out—that's the point of rivers. The running of rivers was a definition of eternity to the Indians of the Six Nations federation. When the English first arrived in Pennsylvania, people from the Six Nations met them with meat, food and animal hides. The people made a treaty with William Penn, promising their friendship "as long as the Sun should shine, or the Waters run in the Rivers." Today, time *is* spoken of as running out, and for all the familiarity of the expression, it betrays an ugly, strange attitude that, for the sake of modernity's exploitative money-making, time may be considered a finite thing. Here by the Taku, time and the river are running out, for the Government of British Columbia in 1998 rushed through approval for the road. (In June 2000, the BC Supreme Court ruled to halt the project, but it could still go ahead as both the Province and Redcorp are appealing the judgment.) Ian Kean, a Canadian who has led a campaign against the road, says "I've got a sense of urgency. In ten years' time, there's not going to be a river like this to protect." The Tlingit peoples with a riverine respect for time and for the Taku, made a poignant statement of their potential loss, describing the "premature," and "speedy" attitude of the white rulers who "fast-tracked" the project, and speaking of the "regulated timeline" for decisions which will rob the "future generations" of Tlingits of their river right. The Tlingits who have lived here for hundreds of years were given forty-eight hours to respond to the decision. What is forty-eight hours to a river?

The Taku represents deep time, glacial time. Forty-eight hours is shallow time indeed. John McPhee writes of the two types of time in *Basin and Range*: "Consider the earth's history as the old measure of the English yard, the distance from the king's nose to the tip of his outstretched hand. One stroke of a nail file on his middle finger erases human history." Composer Peter Maxwell Davies recently wrote of his awe at how "the scale of time changes" in an ice universe when he spent several weeks in Antarctica researching his "Antarctic symphony": "There are endoliths, extremely simple organisms, that live up to ten millimetres down in the coldest Antarctic rock which have the slowest metabolic rate I have ever heard of—they take ten thousand years to transform a minuscule amount of carbon dioxide into proteins. What a lifespan, and what a life!"

The West's dominant attitudes to land and time are—still—the will to enclosure, a desire for private ownership and empire-building. If ever there was a man to epitomize such attitudes of enclosures and colonization, it is Robinson Crusoe. Smugly obsessed with his middle-class status, his preoccupations are enslaving people, enclosing place and scheduling time. After a skirmish with the slave trade, he lands on his island, which he "surveys" and makes his "enclosures" on. Then he turns to time. A chapter title announces priggishly: "I Am Very Seldom Idle." In it, I "order my times of work . . . time of sleep and time of diversion." Ever the colonialist, he colonizes time as much as land, using his calendar to level mountains of months, to carve roads through years, to discipline disorderly time, to honor Christian Sundays and make slaves of savage Fridays. (His name should be Friday, he argues and "I likewise taught him to say 'Master' and then let him know that was to be my name.")

As a precise complementary opposite, the early eighteenth century turned up a Rousseau for a Crusoe. They are radical opposites, over subjects of class and slavery and in their visions of land and of time. Crusoe (in Defoe's book published in 1719) looked at the land of "his" island, setting up fences and enclosures and establishing ownership. Rousseau (born in 1712) detested the philosophy of private land ownership and all acts of enclosure; talking witheringly of the first man who enclosed a piece of land **343** as his own. If only, said Rousseau, someone had pulled up the stakes and said the earth belongs to no one. Rousseau had, like Crusoe, a profound experience of time on an island, but whereas Rousseau writes of the beauty of time by rivers and his understanding of unfettered, wild time, thus: "following the impulse of each instant without any order," and (man is born free, yet everywhere is in chains) refused to wear the metaphysical manacle of the watch, Crusoe was obsessed with clock and calendar-time, with Western, colonial, patriarchal, coercive time, and with shackling every day in a chain of schedule. Rousseau, be it said, is perversely contrary. Though all his beliefs are opposed to Crusoe's, the one book he thought children should have as a library is, well, *Robinson Crusoe*, for Rousseau envied Crusoe's autonomy. A watchmaker's son from Calvin's Geneva, he nonetheless understood the fetter of the watch. Having abandoned his own children, he devised a humane system of education for those of other people. "The most important rule of all education is not to gain time but to lose it."

The desire to enclose common time and to colonize it runs through the whole mind-set of Western time, past and present, and has been one of the main themes of this book. To recap, briefly. Modernity's obsessive time-measurement, noting its passing in minutes, seconds and split seconds, encloses time, metaphorically, with grids and fences. For hundreds of years, Christianity has cornered cyclic time, with its straight-line historicism, squaring the circle of time and enclosing the night. It was deeply implicated in the spread of invisible imperialism, the empire of time: for just as empire builders looked at land and saw *Terra Nullius*, so the metaphysical empire builders looked at time and saw *Tempus Nullius*. Captain Cook was "first to discover" New Zealand, we are told, as if there were no time before White time. (Watch another bit of it. The first European contact with people in Brazil happened on April 22, 1500, at 10 o'clock. Done. *Tempus Nullius* has been clocked.) The history of chronometers is, in effect, the history of colonialism, for once the wild time at sea was logged and tamed, it was used to create the empires of Europe.

In this invisible ideology, those in power in the West have long been colonizing time by defining their time as *the* time—the Puritans, Newtonites, Franklinites and Granthamites—and using their definition as a tool to (their) power; the rich seizing the time of the poor; Europe strangling Africa in slaveries past and present; patriarchal time overruling women's time; Christians colonizing non-Christians; everyone taming children's time. The pretense that "time is money" has profited those who invented the slogan and impoverished everyone else. Wild time, though, was free— the open-handed hour, the open-hearted day, until Franklin and all his banking successors taught that such wild free time was wasted (much as wild land was, equally falsely, called *waste* land).

Unless you are a white, male Christian adult, every celebration of the millennium was an act of metaphysical imperialism, to which a thousand, thousand different cultures must submit. Carnival is intrinsically wild time, but it was colonized, first by the church, then by the class system. Since the Industrial Revolution, both carnival and festival have been tamed, directly by enclosures of the land and indirectly by enclosing its spirit, turning it into pageantry from which the common people are fenced out, excluding the vulgar from their common time, and repressing the vulgar—filthy—wild heart of time.

Genetic engineering, using patent laws as fence posts, steals everyone's common past so that a few multinational "time-owners" can profit from its privatization. In a wider sense, the past is privatized as history, owned and written by the victors, undemocratic and unreachable except through experts. The public past—the Common past—of myth, the uncontainable, diffuse and ambiguous past, democratic and ecocratic, is scorned by modernity and closed off. The future at least ought to be unenclosed, belonging to those who will live there. But it isn't. It has been colonized already, by big businesses who position themselves today to exploit tomorrow. Most ferociously, it is enclosed by the nuclear industry, throwing a radioactive shroud around thousands of years to come. Modernity's dominant attitude to death is to enclose it in hospital walls, to deny its Compost-Heap-Commons and to privatize it for professionals.

Women are fenced out of their own wildest time—their most powerful menstrual times—by the prejudice and hatred of patriarchal societies. Women's faces are "tamed"—to insipidity—by plastic surgery; their bodies' time in childbirth fenced and enclosed by the schedules and timescales of male-time. Though every child is born with a robust inheritance of wild variable time (children can go oh-so-ve-ry-sl-ow-ly, or whizzingly quick), this is disciplined by schooling, tamed to a standard, schooled uniform regularity. Tick. Tock. Tick. Tock.

Art, a natural ally of wild time, with an ability to release the watcher from time-measurement, has been corralled into time-bound definitions. Western art—particularly after the dawn of the global present—has chosen to define itself overwhelmingly in terms of temporal succession. There was futurism and modernism and post-modernism, for every avant-garde, at the *soi-disant* cutting edge, there is a post-garde from the (same) edge. Their sharpest focus is merely the measurement of time. To read some art catalogues is as interesting as listening to the speaking clock.

"Real Art. A New Modernism," is the name of one recent exhibition and the catalogue asks:

Why the apparent tautology, A New Modernism, as opposed to Modernism or Postmodernism? To talk of *A New Modernism* positions it chronologically after Postmodernism, yet presupposes a certain meaning, value and status of Postmodernism, as a discourse that occurred within Modernism as a critique on Modernism but not as the end of Modernism because the Postmodern and the Modern are perpetually constrained in a dialectical relationship with each other, so that the Postmodern succeeds the Modern soley [*sic*] so that it can itself become the Modern: A New Modernism.

Pip pip.

Modern art criticism is, at worst, little more than a clock, end-stopped. Pipped, not *at* the post, but *by* the "post" itself. Most-post-modern. Post-modernist thinking, almost by mistake, magicalizes the moment epipipiphanically, only to undermine it; post-modernism halts time. Post-modern literary criticism shows the last gasp of art's natural tendency to enchant the moment—its fascination with time's high jinks, coincidence, llipsis, pun and *déjà vu vu*. Llusion, both all- and ill-. Yet it is, in the end, a llusion indeed, a game, a maze to time: playful, ludic and losing outside references—the denial of social historical context. Seeking self-reference in time, it creates the still point in the middle of meaning, where time stands still as in a maze-middle. You may have got the gist of the jest—ha ha—you may be amazed at its ingenuity, but it is a lighthearted unenlightenedness, like being in a maze, you have taken a little exercise and got nowhere, for right in this center is an introverted blank, a lacuna, an inverted hedge, a conceptual ha-ha. You get to the sundial in the middle and find it is pointing to nothing but itself, pointing not to noon but to none, to *non*. If a sundial in the maze-middle could point to midnight, here it would, its gnomon gnowing, self-knowing, tipping itself a self-referential wink. And as the atomic clock splits the split second, post-modernism has split the a'om of *belles lettres* and is now splitting airs. Time is slowed to stop, hoist by its own retard. It is a foreknowledge of death in the here and now; not the presence of absence but the prescience of absence.

What hope for wild time? Loads. Every child is born chockful of it, for starters. And then state-of-the-art physics is gloriously telling us about time's chaos, caprice, illogicality and all manner of exuberant disequilibrium. And then there's grown-ups. There's hope in everyone if time can be seen as a—mere—construct. It's hard, though, because since childhood people are taught to see this construct, this edifice of time, as if it had a physical—concrete—quality. Tuesday is spoken of as if it were made of slate, deadline as if it were a wall, "three o'clock" as if a chimney, a minute as if a floorboard and decade as if it were a water tank. We brick ourselves into a house of time which we then find claustrophobic. Perhaps it is precisely because this house of tamed time can become so nastily oppressive that modernity seeks wild time so avidly—demanding wild drugs, wild parties, wild living. But wild time is still everywhere—our measurements of it are not made of concrete but of convention—and we could, if we wanted, huff and puff and blow the house of clock-time down.

So what *is* wild time? What are its characteristics? For answers, look at those things most hostile to the construct of clock-time. Sex, drugs and rock'n'roll. Three ways of having a—vernacular—wild time. As well as these, wild time has a quality of transcendence, of escaping the ordinary, as many religions know. Wild time is profoundly non-Christian, since Christianity has been the most implacable time-tamer. Wild time is the spirit of play, too. All the arts (serious play) have an affinity with wild time, music perhaps most of all. Wild time is nature-based, opposed to the constructed house of clock-time. If modernity's tamed time has a monolithic quality of standardized, global time, one-mono-time:

so the figure who most aptly represents wild time is the mythic Pan, pagan god of nature, exemplifying diversity, as *pan* means "everything," he represents the diversity of wild time itself. The rest of this chapter explores all these characteristics.

Starting with sex. Sex always wants to have its wild way with the hour, shaming the clock. In a beautiful paradox of time, Eros, god of Love, was understood by the Greeks to be at once the oldest and the youngest of the gods, born a child with a chubby wink. Anthropologist Edmund Leach suggests that for the Greeks, "it is the sexual act itself which provides the primary image of time." The clock is the opposite of the erotic. Love detests the clock.

> *Busie old foole, unruly Sunne,*
> *. . . Must to thy motions lovers seasons run?*
> *. . . Love, all alike, no season knowes nor clyme,*
> *Nor houres, dayes, moneths, which are the rags of time, . . .*

as John Donne wrote. It is the clock striking midnight which abruptly stops the burgeoning love affair between Cinderella and Prince Charming.

Lovers have no sense of time and, equally, nothing kills passion faster than noticing the time. As that naughty nursery rhyme knows, and tells with such saucy honesty; the rooster, like today's alarm clock, withers the morning glory:

> *Cock a doodle do!*
> *My dame has lost her shoe;*
> *My master's lost his fiddle stick;*
> *And they don't know what to do.*

And then there's drink. "It is essential to be drunk all the time. That is all: there's no other problem. If you do not want to be the martyred slaves of Time, get drunk, always get drunk! With wine, with poetry or with being good. As you please," wrote Baudelaire. Intoxicants offer a release from tamed clock-time into wild time and surely one reason alcohol and drugs are so much in demand in our over-clocked modern life is that they twist and tangle time, stretch and bounce it and resist the clock's coercion. Perhaps, sadly, this is partly why Aboriginal Australians and indigenous peoples across the world, who have been enclosed in reservations and subjected to Western imperialism of both land and time, have (post-colonialism) such a dangerous need for alcohol as a way of releasing themselves metaphysically into the only freedom still available to them—the wild time of alcoholic oblivion.

Wild time represents transcending ordinary time limits. While writing this book, the only comments I came to dread hearing were "Time is Money, isn't it?" (nope) and "Time? There's not enough of it" as if time were *limited* by a corporate enclosure—Time Ltd. But one Indian peasant woman utterly disagreed: "When god made time, he made plenty of it," was her remark. To her, time was unlimited, unfenced and boundless. The West defines many time limits, arguing that a moment, for instance, is the opposite of an age. Many others disagree. Hindi (with several words for the moon) is positively blasé about time limits. *Kal,* for instance, was sent to try the Western mind-set, for it means tomorrow *and* yesterday: far from placing the moment in opposition to the age, *kal* can mean *both* moment and age. When, in Indian languages, the day-before-yesterday and the-day-after-tomorrow are also expressed by the same words, it offers a glimpse of the centrality of the present. *Now* is what matters. Now transcends into wild time.

Wise religions know this and have long sought to go beyond mere clock-time into a sacred eternity where the secular clock is stopped. (Secular, in its etymology, comes from *saeculum*—to do with time, generation or age.) Shamanistic ecstasy is an attempt to enter a primordial Great Time. Many Eastern teachings emphasize the transcendence of time. Yogis meditate on the instant: the purpose of this is to seek the eternal. Hui-neng, the Sixth Zen patriarch says: "The absolute tranquillity is the present moment. Though it is at this moment, there is *no limit* to this moment, and herein is eternal delight." Zen Buddhism aims to live entirely in the present, with a stance of mindfulness, giving every act full attention, to see the fullness of life even in its most mundane moment. In Hindu thought, the present—and only the present—opens out into the wild time of the unfenced eternal.

While most spiritual disciplines seek to transcend time and to get beyond the clock, in Christianity, clocks, iconically privileged, are put on churches. A clock would look odd on a Buddhist temple. The thirteenth-century mystic Meister Eckhart—unusually for the Christian tradition—speaks of time's uncounted beauty of now; in moments of inner quiet, he said, "There exists only the present instant . . . a Now which always and without end is itself new." The *Tao Te Ching* says: "Move with the present," this present moment which is described as a river in which one must "Go with the flow." (Here, as ever, there is that deep connection between rivers and time.) Modern psychotherapy slightly patly repackages this age-old wisdom as Be Here And Now. And as Seneca asked, in simplicity: "When shall we live if not now?" To reinhabit the now matters, for now is the only time wild time can be lived. Joyce Grenfell, irresistibly Senecan, put it thus: "There is no such thing as time; there is only the minute. And I'm *innit.*"

Wild time is playful. A sense of play—serious play—in Indian mythology is the deepest energy in creation. In the words of Vandana Shiva: "All existence arises from the play of creation and destruction. The manifestation of this energy is called Nature—*Prakriti* . . . *Prakriti* is also called Lalitha, the Player, because *Lila* or play, as free spontaneous activity, is her nature." *Lila* in Hindu myth, plays in endless cycles of time. (Protestant Christianity, with its devotion to the rigid line of work, gave it short straight shrift.)

If nature is the Player, it makes people play. Rafting down the Taku river, someone had brought a huge water pistol, yellow, pink, orange and green fluorescent plastic. A water pistol made in Legoland. By NASA. The water fights raged, *Homo ludens,* alive and screaming, six-year-old Bismarcks with a world to play for. Johan Huizinga, author of *Homo Ludens:* "Mankind at Play" (and a name to play with if ever there wheezinger was one), would have approved. Play is, he argues, a "stepping outside ordinary time." It is an interlude. Culture itself "arises in the form of play." "Play cannot be denied. You can deny, if you like, nearly all abstractions: justice, beauty, truth, goodness, mind, God. You can deny seriousness, but not play." (How does wild time most manifest itself in children? In play.) An experimental psychologist, studying the mind's inner timekeeping, comments "When people are enjoying themselves, they don't pay any attention to their internal clock." (Common enough knowledge: time flies when you're having fun.) And in Nietzsche's words: "All pleasure desires eternity—deep, deep eternity."

Jung remarked: "Civilizations at their most complete moments always brought out in man his instinct to play and made it more inventive." The Tlingit of the Taku have a tradition of the potlatch or gift ritual, a form of play of which Huizinga writes: "The potlatch among Tlingit and others is . . . the most highly developed and explicit form of a fundamental human need, which I would call playing for honour and glory." In *Travels in Alaska,* John Muir writes that after the summer's work is done, "the Takus" as he calls them, "devote themselves to feasting, dancing, and hootchenoo drinking"—hootchenoo being strong spirits and the word thought by some to be the origin of "hooch."

Play matters because, against the gray backdrop of a jobbing sky, play is the rainbow, is energy, is wicked flirtatiousness, is the helplessly laughing, the leglessly laddered, the God of Things which Brimmeth Over, the pint down the pub, the *de trop* overflow of excess, the resplendently unnecessary and the one-too-many which make the whole damn thing worthwhile. Play is *harvest*, is abundance, is generosity, the harvest of pleasure after work, the excess and the gusto, the more-than-enough, the gifts, the spirit of exchange. To play a game is, in German *ein Spiel spielen*, and the brimmingness, the spilling-over abundance of play is mirrored in this brimming-over phrase: spill it, *spiel* it twice— just for fun.

Just for fun, too, in a huge harvest-ready field, the world's biggest maze (a "stonking amount of maze," according to its designer) was made in a *maize* field. The pun was intentional, a play on words, a play on earth, a happy *jeu d'esprit*, and a play on the spirit of harvest. Crop circles, too, are a playful public art in harvest fields. In the same spirit, take the word *giggling*. A one-word harvest of play's superfluity, its liquid, lovely over-indulgence, it has g's to spare (g, the funniest consonant. You want proof? Gnu. Gneed I say more?) and it fills the gaps with "i"—the quickest, wittiest, lickspittiest, trippiest and lightesthearted of all the vowels. Or take the playful digression, the sign of plenty in the barns of the human mind. Or take the idling doodle, so resplendently useless, as exuberantly pointless as the periwig and the peripatetic curlicue, doodles are at once spontaneous and dawdling decoration, doodle for doodle's sake. Sheer play.

Westernized society (urbanized, denatured and ferociously work-oriented) is scared of this panplayism, frightened of its subversive, anarchic nature. Play is too often considered a puerility to be passed ("playtime" is not for grown-ups), a subject to be studied (there is a professor of leisure studies at Pennsylvania State University), a product to be sold (the oxymoronic Leisure *Industry*), or a problem to be solved (the devil finds work for idle hands). Grown-ups resent the Peter Pan in every child and, in the U.K., the government is issuing guidelines that children as small as four must do homework daily. Traditionally, many indigenous peoples do not work for more than four hours a day, which is also the length of time Bertrand Russell suggested in *In Praise of Idleness*; there would then be neither over-employment nor under-employment. He also argues that "there is far too much work done in the world, that immense harm is caused by the belief that work is virtuous." Leisure, by contrast, "is essential to civilisation." The play ethic is far more, well, *ethical* than the work ethic.

Aristotle said, "Nature requires us not only to be able to work well but also to idle well." Sigh. It's so hard. The Idea of the Idle which disses and rejects the clock is an ambition harder than it seems. To really play is to let go of the hand of the clock, to dive deep into the fathoms of time—a state of water this, deep play, with affinities to music, art, sex, deep drinking and deep thought. It is a chancey, risky, fluxy, underwater world where immersion in the moment is all. This wild time is far richer, though far flukier, than clock-time, this is time enlivened and various, time as fast and slow as a waterfall's cascade. It is not necessarily easy to be in, for its waters are uncontrolled by a clock, uncommanded and uncharted. Without a clock you are on your own and it is a difficult but rich experience, this, the beautiful duress of ludic creativity—*idlesse oblige.*

Enough of play. How else is wild time characterized? Music, so connected to time, knows its watery world well. Trance music of tribal peoples is almost always designed to take people out of ordinary time, into an extraordinary, wild, trance-time, for hours or even days. Gamelan music of Java reflects time's eternity and the unchanging models of time central to Java's ancient Hindu-Buddhist beliefs. Java is now predominantly Islamic, with its linear time, and, interestingly, Islamic leaders are suspicious of the profound, nonverbal time-models of gamelan. Gamelan music provokes a division between generations: to Javanese young people, it can seem nondynamic and dull. To the traditional musician, though, youth culture with its linear, goal-oriented music can seem harassing and aggressive.

Time is also important to the Western musical tradition: so in Bach, time is patterned to perfection, in Beethoven it is narrated, in Gregorian chants, time is set in stone awe; but something really interesting happened during the late nineteenth and early twentieth century. Debussy. When I was a child, my piano teacher once played me "The Submerged Cathedral" by Debussy—whose very name sounds like the water his music describes. I was choked. He makes the very notes of time, right *now,* sound from under the pianist's hands. His music is full of sea and full of time—the two as ever intimately connected. Here, the cathedral is a rock or cave under the sea; its stone has sunk as stone must, while the swelling and sinking of the deep gray ocean through its empty gothic windows and arches and pillars make the cathedral resound, ringing as if it were itself a musical instrument rung and sounded into chords. And this music can still make me cry, the body's saltwater crying for the saltwater of the unsounded ocean's sound.

All forms of art happen in an atemporal state: the disposition of absorption, solely in the now, when the clock's tyranny is deposed. It is a truism to note how artists of all sorts lose track of time while working, but this sense of moment can be, as it was with Debussy, the revelation of their work.

As the first chapter hinted at, just when time had been standardized, just when there was first a sense of global time and when wildernesses and humankind were just about to reach their critical moment and change places, just when, in short, time had never seemed so tamed was exactly when the most brilliant musicians in the Western tradition exploded with a rendition of wild time. When Mahler wrote music that passes through time into silence, through sound into wild time, he knew with his whole soul what wild time's eternity was for, and how to touch it. In 1894, Debussy wrote "Prélude à l'après-midi d'un faune," with its unforgettable, wild flute opening, which falls right off the hill-sides, barrows and woods. Modern music, said Pierre Boulez, "awakes" with the flute of Debussy's *faune*. Like a wilderness with its wild and waking grandeur, the waking flute or pan-pipes are highly significant; Pan himself is the piper at dawn, who, by tra-dition, like the Pied Piper, plays you out of towns and roads (those concrete expressions of linear time) out into a green wilderness to waltz on the wild side of time.

In the two years after the prelude, Mahler wrote his third symphony, originally called the Pan symphony. This music, pas-sionate, playful and wild, nonetheless hums with erotic intensity which would burn and bruise soon as look at you; it drums the wild and earthiest drumming. Eyelids lowered, it evokes the smoky moment with the dark impatience of its erotic rhythms. Stravinsky's "Le Sacre du Printemps"—"The Rite of Spring"—was a pagan fertility rite, which he described as "*une sorte de cri de Pan.*" Pan, whose cry was heard only by musicians (those with the sharpest ears), was perhaps the only cultural icon the West had which was powerful enough to oppose the dominant coloniza-tion and taming of time. Pan is wild.

Pan is the pagan god of nature, of wild place, the original wilderbeastie, creature of wild woods and wilder ways; he per-sonifies wilderness and—to me—wild time. Pan, earth god, plays on the pipes named after him. (Earth music is traditionally played on pipes; panpipes, bagpipes, didgeridoos or shepherd's pipes.) He represents all the qualities of wild time, in art, in play, in tran-scendence, in intoxication and in the sexual heart of life. With Christ's crucifixion, the Christian church announced "Great Pan is Dead" and this echoed across the world, but it was never so. Pan, taking a running jump across centuries, lands, heels making a deep impression in the mud, with a wild and gap-toothed guffaw and a raging hard-on, and his cry, echoed in the music of Stravinsky and Debussy, is still audible in the resurgence of paganism today.

The pagan calendar dislikes homogeneity, disputes the Christian influence on the standard calendar and despises the time (at once global *and* suburban) of industry and finance. The pagan calendar variegates time, using nature, in the form of moons, solstices and equinoxes, as a marker of common time. Thus a full moon is a public festival, and each full moon has a different name; the barley moon, mead moon, wolf moon and harvest moon. The pagan calendar characterizes time, cherishing difference; it responds to the rhythm of the seasons, not the workplace; its time is lunar not consumer, its holidays decreed by sun and moon, not governments and banks. On the first of May, it has Beltane, the spring celebration of fertility—the potential of time. Lughnasadh, at midsummer, celebrates harvests and the fullness of time. Samhain, the festival acknowledging death, reveals the ending of time. Imbolc, the pagan midwinter festival, tells of the waiting of time, to be reborn at Beltane. In the pagan calendar, time is animated, enchanted and characterized.

Pan is celebrated, reincarnated in a contemporary Beltane bash, a ripsnorter of a festival with fire-breathers, fire-jugglers, the firing of an earth maze and a hundred puckish figures dancing naked around a bonfire to panpipes and drums; fireleaping to the music of bagpipes, didgeridoos and a yipyipyip chirrup, intoxicated and intoxicating. There is something Bacchic here, the pan-abundance, like the spirit of play, something of the seven-too-many and sleep it off for a week.

Pan was Public Enemy Number One to the Christian church. While he stood for all the aspects of wild time, Christianity stood for its antithesis, the taming of wild times, in all ways. Pan, horny beast with fuck-anything eyes, is, of course charged with lust for sex. Pan, pure nature, fired by impish and impious impropriety is deliciously impure in thought, word and deed. Pan, half-man, half-goat and all bollocks; Pan, with two pointy ears and one pointy prick; Pan, shaggy-thighed and rooting for a shag in every tree root, was called Auld Hornie by Christian mythologists, who, in the Westernized world, replaced his filthy fecundity with a Virgin Mother, a purse-lipped "clinliniss is nixt to godliniss," and a two-thousand-year hatred of the profoundly sexual, cyclic and procreative nature of time. Pagans across the world celebrated sex with tall stiff poles planted in the round wet hole of mother earth, dancing around it for fertility. The Christian authorities fully understood the Maypole's dirty dancing "obscenity," for if paganism lip-lickingly says Yes to sex, Christianity, tight-lipped, says No. Going all fig-leafy, the Christian church wanted great Pan dead. But Pan is panoramically alive, life in lust for life; Pan is in the panplayism of party animals, Pan is the spirit of wild time; feel the hot breath down the back of your neck, feel the thump of heel on mud, the panache of time set free, roguish and ribald, in a rank and reckless, rocks-off, bollock-naked uncorked stomp.

Such wild time is implacably non-Christian. It's worth pausing to look at this in detail, because the Church has been breathtakingly influential on the Western idea of time, from the history of clocks to time-use and the taming of wild time. Christianity hates and fears all wild time's manifestations: sex, drugs and the play ethic, trance, dance and nature's time. Hence, as previous chapters show, Christianity's long love-affair with the clock.

Start with sex. Christianity is an enemy of sex. "Better to marry than to burn with lust," spits St. Paul, but that's as good as it gets. Sex and the clock, though, hate each other. Operating on the principle that my enemy's enemy is my friend, Christianity found more and more reasons to befriend the clock, especially clocks and bells which counted through the night, when sex most marauded. Sundials, for instance, only fenced and tamed daytime and there could be no "hours" at night, so, after twilight, as the sloping finger slooped down, time sloped off into the shadows of dark evening, of wild time. Christian monks—celibate, one notes—were the first to divide the hours of the night, fencing its commons, interrupting untold acts of coitus with every disapproving peal of bells, each speaking of the dominion of Christian, monastic values.

Welsh poet Dafydd ap Gwilym, in the second half of the fourteenth century, wrote the following attack on clocks. While staying a night in a monastery, the poet dreams an amorous dream of his woman in a distant town. A clock wakes him, shattering his dream, mentally forcing him back to the monastery and its *mores*.

Woe to the dark-faced clock beside the hedge
That woke me up.
Useless be its head and its tongue
And its two ropes and its wheel
And its weights, those blunt-shaped balls,
And its yards and its hammer
And its ducks thinking it's day
And its restless mills.
Useless clock like a crazy click
Of a drunken cobbler, cursed be its shape. . . .
A goblin's mill grinding the night.
Was a saddler or a crupper of scabs
Or a roof-tiler more unsteady?
Cold destruction take its cry
For leading me here from heaven!

Six hundred years on, you can still hear him gnashing his teeth.

While wild time is rawly alive (the spirit of Pan erupting in boiling vigor from barrows and forests, orgiastic and effervescent—*I am alive*): Christianity's sense of time is death-related, the

Christian cross represents the dying god on a dead tree. Christianity loves its relics, mausoleums and Holy Ghosts, the communion wafer and wine is the symbol of dead white flesh and dying blood. Christianity focuses not on this life but on the life hereafter.

In its crusade against wild time, Christianity also battled against wild licentiousness—off-license time, boiling down flagons to a sip and cutting down loaves to a wafer—thin as a sneer. The Temperance movement and Victorian Christian morality put fences of clock-time around the wild time of the hophead and the sot—strictures of laws which have lasted—in Britain—ever since. See it today, in the enclosures of "closing time," fences or bars in bars and pubs across the country when a bell—straight out of the Benedictine orders—rings a great big *DONG,* meaning "TIME at the bar, now, please," and common people are excluded, barred once again from their common—wild—time. (The rich, with their private drinking clubs, need obey no such laws.)

Idleness was the enemy of the soul to Benedictines. Play was arch-idleness, and Protestant Christianity damned it with a one-way ticket to hell on a monorail. (Says Huizinga, "Calvin and Luther could not abide the tone in which the Humanist Erasmus spoke of holy things. Erasmus! his whole being seems to radiate the play-spirit . . . . his light irony and adorable jocosity.") The church, of course, invented the Protestant work ethic when, as part of the Industrial Revolution, capitalists and industrialists made a deal with god in the boardroom. Christian missionaries outlawed the Tlingit potlatch, highest form of play, in their worldwide drive to destroy all wild time. Missionaries banned trance music, dance music and carnival-time from Burma to Brazil and told indigenous peoples that their wild times belonged to the devil. John Muir describes the Stickeen Indians dancing and then performing individual animal dances, the bear dance, porpoise and deer dances. "These animal plays were followed by serious speeches, interpreted by an Indian woman: 'Dear Brothers and Sisters, this is the way we used to dance. We liked it long ago when we were blind, we always danced this way, but now we are not blind. The Good Lord has taken pity upon us and sent his son, Jesus Christ, to tell us what to do. We have danced to-day only to show you how blind we were to like to dance in this foolish way. We will not dance any more.'"

And what did the church think of giggling? A mandate from the Bishop of Exeter in 1330 was issued against giggling in church: "shameful to relate and horrible to hear," said the Bishop, were these "laughings, gigglings and other breaches of discipline."

Wild time is nature's time: and if ever a religion hated nature, it is Christianity. As Christianity made its onslaughts on the pagan world, pantheism and the spirit of Pan endured in nature's deepest darkest places, in forests and wildernesses, not in towns. Grooving the sacred groves (all sexual symbolism fully intended), paganism lasted longest in the woods where oak gods were honored, trees were dressed and wood demons appeased. Christianity was spread by roving missionaries literally making "inroads" into pagan forest areas. Pagan, etymologically, means "of the countryside." Christianity, in cultural opposition, is on the side of *roads.* Christ is the Way, or Road. St. Paul's most significant moment was on the *road* to Damascus. St. Augustine wrote *The **City** of God.* Paved, one presumes. Christianity loved the straight Roman roads and used them for proselytizing and for social control; using roads to kill wild nature's pagan past. Roads are still killing nature today, from tearing up wildernesses (witness the invidious road proposed in the Taku area) to encouraging traffic, and greenhouse gases. In Britain, hundreds of the pro-nature, anti-road protesters are pagans (one calling himself Pan), all celebrating Beltane and living, as they call it, in Rainbow Time. Christianity, deeply suspicious of the untamed, be it wild land, or wild mind, drove a road of office hours through the wild time of night and organized history into rigid white motorway lines.

Christian architecture—of both space and time—is phallic, straight and linear. Square-churched Christianity commands its congregations to sit in rows of pews and brought people indoors to tame their time. Christianity loves its plumblines, from church clock-towers to Freemasonry's symbolism. History was made linear and wild time was cornered by Christianity and by all its straight-line adherents. Nature's architecture, by contrast, again of both space and time, from the circle of the year to the circling of a snail's shell; is round as a nest (native peoples meet in talking feather "circles" or council circles, pagans circle dance), it curves to the influence of the circle principle, the vaginal principle. Mother Nature is female and she abhors a plumb line.

Pan as prefix means "all" or "everything." In Pan the goat-god, it symbolizes the multifold animism of pantheism. Diversity and difference. Christianity went for the mono—the one, monotheistic, monocultural—the uniform monotonous same. Christianity, the tarmac of religions, makes uniform what was once diverse. One-god, one-world Christianity has a monocul-turalizing influence; destroying the rippling variety of spirits of the land and site-specific animism to one leveled tarmac. Jesus Christ, like suburbia, is the same yesterday, today and forever. Time is flatly predictable; the Christian god, always the same, all codified and good as his word, is not your man for a tantrum. Pan, unguessable and self-pleasuring, is king of the mood-swingers, time's capricious unpredictability like a dissipative structure at its bifurcation point. The mono-principle of Christianity has influenced so much of modernity's monoistic mind-set, whether the monolithic corporations, monolithic time zones of international trade, mono-crops or the mono-definition of time itself (*the* time). If Christianity influences, and represents, "mono" time,

Pan represents all time's variety; the biodiversity of time and, in a wilderness above all, time is—still—wild.

In wildernesses, time is as diverse as the play of light across a year of landscape. Just as the human need for wilderness becomes more acute with increasing development and a cockroaching population; so our need for wild time gets greater as the encroaching clock shunts its way across the mind. If we lose wilderness we lose the visible picture of wild time; the future will never know the time of snow and fire, time which thinks in grandeur like an ancient tree, which moves in passages of stature like a mountain, which knows the long white wait of a waterfall, running so fast and falling so slowly.

Here, the Taku is one of the most seasonally time-full places on earth; in winter all bear-den-dark, while in summer the light stretches itself thin over the top of midnight so you cannot tell if it is the end of the longest of sunfalls or the beginning of the longest of sunrises. In the Taku, you can know the dredge of a primeval age enduring, or the sheer shine of liquid instantaneity in a salmon's leap. Here, time has a variety no clock will ever know or mark, where an eagle's veering flight puts time on pause as it unfurls to hover before you, wind made majestic, time held in a scroll of wings.

Nothing lolls like a bear can loll; here bears on the mooch snuffle huckleberries. Nothing moves quicker than the flash of claw in water as a bear catches fish in its paws. Nothing can compete for the sheer diversity of time; here a five-minute tornado can fell a forest and a scamp of a sudden current can skim a log a mile downriver. And then time can stop on a glacier, leaving the signature of ten thousand years ago to last ten thousand years, written in the sheets of ice of an ancient frozen river, running now at the speed of never. From that glacial time, massive—macro—time you move to the minute—micro—time; in the insect world, an hour is like a season, and a season like a generation.

Red is, symbolically, the color of mortality; blood of life and of death: blue, the color of eternity. Here in this time-diversity, they are found side by side. The blue of a glacier, the red of an—inexplicable—ladybug. Nothing is older than the blue glacier, ten thousand years in the making, ten thousand years in the unmelting. Nothing is younger than the bright red button of a ladybug hatched at the beginning of this sentence: here is the chasmic grandeur of wild time—a ladybug's little red-letter day tickling for a minute the glacial blue ice of eternity.

BIBLIOGRAPHY SOURCES & SUGGESTIONS FOR FURTHER EXPLORATION

Most of the following books were used glancingly for specific chapters and are cited under the chapter headings below. Certain books, though, informed mine more fully and are listed under the General Bibliography.

GENERAL BIBLIOGRAPHY

Adam, Barbara. *Timewatch: The Social Analysis of Time* (Polity Press, 1995).
Important, observant and relevant sociology.

Chesneaux, Jean. *Brave Modern World* (Thames & Hudson, 1992).
This book, particularly in its resplendent opening, fizzes.

Forman, Frieda Johles & Caoran Sowton. *Taking Our Time: Feminist Perspectives on Temporality* (Pergamon Press, 1989).
An excellent (and rare) look at the subject of time and women.

Macnaghten, Phil & John Urry. "Nature and Time" in *Contested Nature* (Sage Publications, 1998).
A wise, wide, deep look at nature, culture and time.

Mumford, Lewis. *The Myth of the Machine: Technics and Human Development* (Secker & Warburg, 1967). A trenchant gem of a book.

Priestley, J. B. *Man and Time* (Aldus Books, 1964).
A one-man's-a-wondering, by a great muser on things.

Quinones, Ricardo. *The Renaissance Discovery of Time* (Harvard University Press, 1972).
A piquant, eloquent critic.

Rawlence, Christopher (ed.). *About Time* (Jonathan Cape with Channel Four, 1985).

Imaginative, instinctive and sensitive, unexpected and beautiful, this is (was) TV working at such a pitch of excellence that it both translates into a book and stays current well over a decade after transmission.

Thompson, E. P. "Time, Work Discipline and Industrial Capitalism," *Past and Present,* No. 38 (December, 1967).

Funny and erudite, political and literary, sweeping and detailed, this is a compact, brilliant essay on the Industrial Revolution and time.

Whitrow, G. J. *What Is Time?* (Thames & Hudson, 1972).

Extraordinarily careful and detailed account of the history of time.

The following three were useful and pleasurable to read; facts from their historical research found their way into a number of chapters:

Aveni, Anthony F. *Empires of Time: Calendars, Clocks and Cultures* (I. B. Tauris, 1990).

Cipolla, Carlo M. *Clocks and Culture 1300–1700* (Collins, 1967).

Landes, David. *Revolution in Time, Clocks and the Making of the Modern World* (Belknap Press, Harvard University Press, 1983).

Two books which are not—it would appear—about time at all, but both of which have it as a strong subtext:

Girardet, Herbert. *Earthrise: Halting the Destruction, Healing the World* (Paladin, 1992).

A sturdy, angry took at nature's and human time.

Lane, John. *A Snake's Tail Full of Ants: Art, Ecology and Consciousness* (Resurgence Books, 1996).

A gracious and elegant read.

And finally, although detailed later, the following writers were an inspiration:

Raimondo Panikkar, Mircea Eliade, Paul Virilio, Ivan Illich, Bob Bushaway, Penelope Shuttle and Peter Redgrove, Vandana Shiva, Fritjof Capra, George Steiner and Johan Huizinga.

1. PIPS AND OCEANS AND THE NOW

Abensur, Nadine. *The New Cranks Recipe Book* (Weidenfeld & Nicolson, 1996).

Annales Cumbriae (1860).

Barbour, Julian. Quoted in *Discover* magazine (December 2000).

Baudrillard, Jean. *The Illusion of the End* (Polity Press, 1994).

BBC Radio 3, 21 May 2001 on jetlag.

Bergson, Henri. Quoted in Priestley's *Man and Time*.

Blake, William. "Auguries of Innocence" (1789).

Byron, George Gordon. *Childe Harold's Pilgrimage* (1812–1818).

Cottle, T.J. and S. L. Klineberg, *The Present of Things Future* (Free Press N.Y., 1974).

Davis, Wade. Quoted in *The World and the Wild*, (ed.) D. Rothenberg and M. Ulvaeus (University of Arizona Press, 2001).

Dogen. *Time-Being.*

Dogen. *The Moon in a Dewdrop.*

Eliade, Mircea. "Time and Eternity in Indian Thought" in *Man and Time* (ed.) Joseph Campbell (Princeton University Press, 1957).

Evans-Pritchard, E.E. *The Nuer: A Description of the Modes of Livelihood and Political Institutions of a Nilotic People* (Clarendon Press, Oxford, 1940).

Feld, Steven. *Sound and Sentiment: Birds, Weeping, Poetics and Song in Kaluli Expression* (University of Pennsylvania, 1982).

Foster, Craig and Damon (Dirs.). *The Great Dance: A Hunter's Story* (2000).

Frazer, James George. *The Golden Bough: A Study in Comparative Religion* (Macmillan, 1890).

Friedman, Meyer. *Treating Type A Behavior and Your Heart* (Knopf, 1984).

Gaffield and Gaffield. *Consuming Canada: Readings in Environmental History* (Copp Clark, Toronto, 1995).

Gault, Richard. "In and Out of Time," *Environmental Values 4* (The White Horse Press, Cambridge, 1995).

Givens, Douglas R. *An Analysis of Navajo Temporality* (University Press of America, 1977).

Grand Street 59. *Time* (Grand Street Press, 1997).

Gray, Andrew. *The Last Shaman: Change in an Amazonian Community* (Berghahn Books, 1997).

Ha, Tae Hung. *Folk Customs and Family Life* (Yonsei University Press, Seoul, Korea, 1958).

Hallmark Cards, Inc. *1994 Date Book.*

Harding, Michael. "Astrology: The Language of Time," *Journal of the Society for Existential Analysis,* Vol. 4.

Hardy, Thomas. *Tess of the D'Urbervilles* (1891).

Holub, Miroslav. *The Dimension of the Present Moment* (ed.) David Young (Faber & Faber, 1990).

International Herald Tribune, 22 March 2001 on Boulder, Colorado.

Joyce, James. *Ulysses* (Shakespeare & Co.; Paris, 1922).

———. *Finnegans Wake* (Faber & Faber, 1939).

Kane, Joe. *Savages* (Macmillan, 1995).

Kastan, David Scott. *Shakespeare and the Shapes of Time* (Macmillan, 1982).

Kern, Stephen. *The Culture of Time and Space 1880–1918* (Weidenfeld & Nicolson, 1983).

Krolick, Sanford. *Recollective Resolve: A Phenomenological Understanding of Time and Myth* (Mercer University Press, 1987).

Leopold, Aldo. *A Sand County Almanac* (Oxford University Press, 1987).

Levine, Robert. *A Geography of Time* (Basic Books, 1998).

Lipsitz, George. *Time Passages: Collective Memory and American Popular Culture* (University of Minnesota Press, 1990).

Millar, Jeremy. "Rejectamenta" in *Speed: Visions of an Accelerated Age* (ed.) Jeremy Millar & Michiel Schwarz (The Photographers' Gallery & the Trustees of the Whitechapel Art Gallery, 1998).

de Montaigne, Michel quoted in Quinones' *The Renaissance Discovery of Time.*

Narayan, R. K. On childhood in the *Guardian* (October 1995).

New York Post, 12 September 2001.

The New York Times, "Science Times," 19 June 2001.

Östör, Ákos. *Vessels of Time: An Essay on Temporal Change and Social Transformation* (Oxford University Press, 1993).

Ovid, *Metamorphoses.*

Panikkar, Raimundo. "Time and History in the Tradition of India: Kala and Karma" in *Cultures and Time* (ed.) L. Gandet et al. (UNESCO Press, 1976).

Peat, F. David. *Blackfoot Physics: A Journey into the Native American Universe* (Fourth Estate, 1995).

Piaget, Jean. *The Child's Conception of Time* (Routledge, 1969).

Posey, Darrell Addison. On the Kayapo or Mebengokre, in *Resurgence* (Nov./Dec. 2000).

Poulet, Georges. *Studies in Human Time* (Johns Hopkins Press, 1956).

Pritchard, Evan T. *No Word for Time: The Way of the Algonquin People* (Council Oak Books, 1997).

Proust, Marcel. *À La Recherche du Temps Perdu* (Bernard Grasset, Paris, 1913).

Reichard, Gladys. *Navaho Religion: A Study of Symbolism* (Pantheon Books, 1950).

Reichel-Dolmatoff, Gerardo. *The Forest Within* (Themis Books, 1996).

———. *The Shaman and the Jaguar* (Temple University Press, 1975).

Rosaldo, Renato. "Ilongot Visiting: Social Grace and the Rhythms of Everyday Life," published in Creativity/Anthropology (eds.) S. Lane, K. Narayan and R. Rosaldo, (Cornell University Press, 1993).

Rousseau, Jean-Jacques. *The Reveries of the Solitary Walker* (1782).

Service, Elman. *The Hunters* (Englewood Cliffs, N.J., 1966).

Shakespeare, William. *Twelfth Night.*

———. *As You Like It.*

Sobel, Dava. *Longitude* (Fourth Estate, 1996).

Telecom, British "Rugby Radio Time and Frequency" (National Physical Laboratory, 1997).

Thomas, Dylan. *Collected Poems* (Dent, 1952).

Thoreau, Henry David. *Journal* January 26, 1852, Neufeldt and Simmons, (eds.) (Princeton University Press, 1992).

———. *Works* (Princeton University Press, 1971).

Wacziarg, F. and A. Nath, *Rajasthan: The Painted Walls of Shekhavati* (Croom Helm, 1982).

Wittgenstein, Ludwig. *Notebooks 1914–1916* (Chicago, 1979).

Wokler, Robert. *Rousseau* (Oxford University Press, 1995).

Woolf, Virginia. *The Waves* (L. and V. Woolf, 1931).

Zerzan, John. *Elements of Refusal* (Left Bank Books, 1988).

2. F.FWD. THE TROUSER-ARROW OF SPEED

Basso, Keith H. *Southwestern Journal of Anthropology,* Vol. 26, No. 3 (Autumn 1970).

Bertmann, Stephen. *Hyperculture: The Human Cost of Speed* (Praeger, 1998).

Bitomsky, Hartmut. Quoted in Edward Dimendberg's "The Will to Motorisation" in *Speed. Visions of an Accelerated Age,* (eds.) Jeremy Millar and Michiel Schwarz.

Bourdieu, Pierre. *Mediterranean Countrymen* (Paris, 1963). Quoted in Thompson's "Time, Work Discipline and Industrial Capitalism."

Carroll, Lewis. *Alice's Adventures in Wonderland* (1865).

Carson, Rachel. *Silent Spring* (Hamish Hamilton, 1963).

Chatwin, Bruce. *The Songlines* (Jonathan Cape, 1987).

Control, Mr. Social. *Away with All Cars* (Playtime For Ever Press, 1992).

Dallas Times Herald, 25 March 1991, on drive-thru funeral parlors.

Donnelley, Paul. *Max Power: Top Mad Motors, Top Totty, Top Ice, Top Cruises, Top Speed, Top Book* (Andre Deutsch, 1998).

Evening Standard, 4 December 2000, on Oxford Street.

George, Susan. "Fast Castes," in *Speed: Visions of an Accelerated Age,* (eds.) Jeremy Millar and Michiel Schwarz.

Goldman, Ari. L. "Religion Notes" *The New York Times,* 12 February 1994.

Gunston, B., D. Taylor, & A. Ewart (eds.). *The Guinness Book of Speed: Facts and Feats* (Guinness Superlatives, 1984).

"How America Has Run Out of Time," *Time,* 24 April 1989.

IH8U: ltle bk of txt abuse (Michael O'Mara Books Limited, 2001).

Illich, Ivan D. *Energy and Equity* (Calder & Boyars, 1974).

International Herald Tribune, 9 March 2001, on children's time.

————, 23 February 2001, on Japan and politeness.

————, 3 January 2001, on speed.

Kundera, Milan. *Slowness* (Faber & Faber, 1996).

Lee, Laurie. *As I Walked Out One Midsummer Morning* (Andre Deutsch, 1969).

Marinetti, Emilio. *Manifesto of Futurism* (1909).

————. *The New Religion-Morality of Speed* (1916).

Maybury-Lewis, David. *Millennium* (Viking, 1992).

McKibben, Bill. *The End of Nature* (Penguin, 1990).

McKie, Robin. *On Punkin Chunkin in the Observer* (26 October 1997).

Montague, Peter. "The Obscenity of Accelerated Child-Development," in *The Ecologist* Vol. 28, No. 3 (May/June, 1998).

The New York Times, 13 November 1996, quoting Klaus Schwab.

O'Driscoll, Kieran & Declan Meagher. *Active Management of Labour* (Mosby, 1993).

Orr, David W. "Slow Knowledge," *Conservation Biology* Vol. 10, No. 3 (June, 1996).

Poggioli, Renato. *The Theory of the Avant-Garde* (Belknap Press, Harvard University Press, 1968).

Sachs, Wolfgang. "The Speed Merchants," *Resurgence,* No. 186 (January/February, 1998).

————. "Speed Limits," in *Speed; Visions of an Accelerated Age,* (eds.) Jeremy Millar and Michiel Schwarz.

Sun Tzu (joint authors). *The Chinese Art of War.*

Swiss government research on "battery children." Quoted in the *Observer* (29 March 1998).

Virilio, Paul. *L'Horizon Négatif* (Galilée, 1984).

————. "Speed and Politics: Essay on Dromology" *N. Y. Semiotext(e)* (1986).

Williams, Heathcote. *Autogeddon* (Jonathan Cape, 1991).

Whitelegg, John. "Time Pollution," in *Transport for a Sustainable Future* (Belhaven Press, 1993).

————. *Transport Policy in the EEC* (Routledge, 1988).

————. *Urban Transport* (Macmillan Education, 1985).

————. *High Speed Trains* (Leading Edge with Stockholm School of Economics, 1993).

————. *The Spirit and Purpose of Transport Geography* (University of Lancaster, 1981).

Worsley, Peter. *Knowledges* (Profile Books, 1997).

3. MYTHICAL LIZARDS, MARS BARS AND ASPIDISTRAS OF THE PAST

Allen, Louis A. *Time Before Morning: Art and Myth of Australian Aborigines* (Rigby, 1976).

Barthes, Roland. *Mythologies* (Jonathan Cape, 1972).

Basso, Keith. "Stalking with Stories: Names, Places and Moral Narratives Among the Western Apache," in *Text, Play, and Story: The Construction and Reconstruction of Self and Society* (Proceedings of the American Ethnological Society, 1983).

Bettelheim, Bruno. *The Uses of Enchantment* (Thames & Hudson, 1976).

The Bible.

Braid, Mary. On David Coulson, in the *Independent on Sunday* (21 September 1997).

Brody, Hugh. *Maps and Dreams* (Faber, 1981).

————. *The Other Side of Eden: Hunter-Gatherers, Farmers and the Shaping of the World* (Faber, 2001).

Campbell, Beatrix. "The Autumn of the Matriarch," in the *Guardian* (15 March 1997).

Campbell, Joseph. *Myths to Live By* (Paladin Books, 1985).

Collingwood, R. G. *The Idea of History* (Clarendon Press, Oxford, 1946).

Cowan, James. *Letters from a Wild State: An Aboriginal Perspective* (Element Books, 1991).

————. *Mysteries of the Dream-Time: The Spiritual Life of Australian Aborigines* (Prism Press, 1992).

Das, Rabindra Kumar. *Mysterious Konarka* (Kitab Mahal, India, 1984).

Descola, Philippe. *The Spears of Twilight: Life and Death in the Amazon Jungle* (HarperCollins, 1996).

Drillbits and Tailings: *www.moles.org/ProjectUnderground/drillbits.*

Dunne, John S. *Time and Myth* (University of Notre Dame Press, 1975).

Durning, Alan Thein. *Worldwatch,* Paper 112.

Eco, Umberto. *Travels in Hyperreality* (Pan, 1986).

The Ecologist, Vol. 31, No. 4, May 2001, on the Jabiluka mine.

Eliade, Mircea. *Rites and Symbols of Initiation* (Harper & Row, 1965).

————. *The Sacred and the Profane* (Harper & Row, 1961).

————. *The Myth of the Eternal Return* (Routledge & Kegan Paul, 1955).

————. *Myth and Reality* (George Allen & Unwin, 1964).

————. *Myths, Dreams and Mysteries* (Harvill Press, 1960).

————. *Images and Symbols* (Harvill Press, 1961).

Flannery, Tim. *Throwim Way Leg: Adventures in the Jungles of New Guinea* (Weidenfeld and Nicolson, 1998).

Fonseca, Isabel. *Bury Me Standing: The Gypsies and their Journey* (Chatto & Windus, 1995).

Gould, Stephen Jay. *Time's Arrow, Time's Cycle: Myth and Metaphor in the Discovery of Geological Time* (Harvard University Press, 1987).

Gray, Andrew. *The Arakmbut: Mythology, Spirituality and History in an Amazonian Community* (Berghahn Books, 1996).

Griffiths, Thomas. *Forest Peoples Programme Report,* August 2000.

Hersey, John. *Hiroshima: An Account of Events Following the Dropping of the Atomic Bomb on Hiroshima, August 6, 1945* (Penguin, 1946).

Hugh-Jones, Christine. *From the Milk River: Spatial and temporal processes in Northwest Amazonia* (Cambridge University Press, 1979).

Independent, 26 May 2001, on Weatherman Draw.

Lowenthal, David. *The Heritage Crusade and the Spoils of History* (Viking, 1997).

Massola, Aldo. *The Aborigines of South Eastern Australia As They Were* (Heinemann, Australia, 1971).

Monbiot, George. *No Man's Land: An Investigative Journey Through Kenya and Tanzania* (Macmillan, 1994).

Morton, Andrew. *Diana: Her True Story* (Michael O'Mara Books 1993).

Mother Jones, January/February 2000, quoting Roberta Blackgoat.

Munn, N.D., in *Australian Aboriginal Anthropology,* Berndt, R.M., ed. (University of Western Australia Press, 1970).

Munz, Peter. *The Shapes of Time: A New Look at the Philosophy of History* (Wesleyan University Press, 1977).

Narby, Jeremy. *La Vision des Autres: Les Amerindiens et la "Decouverte" des Ameriques* (SAVED, 1990).

———. *The Cosmic Serpent: DNA and the Origins of Knowledge* (Victor Gollancz, 1998).

The New York Times, 13 September 2001, on the U.S. flag.

The New Yorker, 20 November 2000, on Jean-Luc Godard.

Rosaldo, Renato. *Ilongot Headhunting: A Study in Society and History* (Stanford University Press, 1980).

Ryle, John. On temples in Bhutan, in the *Guardian* (27 April 1998).

Samuel, Raphael & Paul Thompson. *The Myths We Live By* (Routledge, 1990).

Santos-Granero, Fernando. *Time Is Disease, Suffering and Oblivion: Yanesha Historicity and the Struggle against Temporality.* Paper presented at session "History and Historicity in Amazonia," Annual Meeting of the American Anthropological Association (Chicago, 1999).

Shepard, Paul. *The Others: How Animals Made Us Human* (Island Press, 1996).

Taggart, James M. *Enchanted Maidens: Gender Relations in Spanish Folktales of Courtship and Marriage* (Princeton University Press, 1990).

Telecom, British. *The Phone Book: London Postal Area* (April 1997).

Thornton, Robert J. *Space, Time and Culture Among the Iraqw of Tanzania* (Academic Press, 1980).

Tsing, Anna. *In the Realm of the Diamond Queen: Marginality in an out-of-the-way place* (Princeton University Press, 1993).

Ussher, James. *Annals of the World* (1650).

Whitfield, Stephen J. *American Space Jewish Time: Essays in Modern Culture and Politics* (North Castle Books, 1996).

Wolf, Eric. *Europe and the People Without History* (University of California Press, 1982).

www.atsic.gov.au, on the Jabiluka mine.

www.ivf.com, "Eight is Enough."

Yohannan, John D. (ed.) *A Treasury of Asian Literature* (New American Library, 1958).

4.   BOTTOMS UP! MISCHIEF NIGHTS AND MILLENNIUM DAYS

Anson, Robert Sam. On the Millennium Bug, in the *Observer* (13 December 1998).

Avery, Gillian & Julia Briggs, (eds.). *Children and their Books* (Clarendon Press, Oxford, 1989).

Bauman, Richard (ed.). *Folklore, Cultural Performances and Popular Entertainments* (Oxford University Press, 1992).

Bloom, Harold. *Omens of the Millennium: The Gnosis of Angels, Dreams and Resurrection* (Fourth Estate, 1996).

Bunting, Madeleine. On the Asda harvest festival, in the *Guardian* (11 October 1997).

Bushaway, Bob. *By Rite: Custom, Ceremony and Community in England 1700–1880* (Junction Books, 1982).

Campbell, Alan Tormaid. *Getting to Know Waiwai: An Amazonian Ethnography* (Routledge, 1995).

Cohn, Norman. *The Pursuit of the Millennium* (Pimlico, 1993).

Defoe, Daniel. *A Tour Through the Whole Island of Great Britain* (1724–6).

Eco, Umberto. *Apocalypse Postponed* (Flamingo, 1995).

Groening, Matt. *The Simpsons,* episode one: "Simpsons Roasting on an Open Fire." (first aired 17 December, 1989).

Hare, David. Quoted in the *Guardian* (25 February 1998).

Heschel, Abraham Joshua. *The Sabbath* (Farrar, Straus & Giroux, 1951).

Kinser, Samuel. *Carnival, American Style: Mardi Gras at New Orleans and Mobile* (University of Chicago Press, 1990).

Knight, Richard. *The Millennium Guide* (Trailblazer Publications, 1998).

Liverpool Echo. Article on Asda harvest festival (23 September 1997).

Massingham, H. J. "William Shakespeare of Warwickshire" in *Where Man Belongs* (Collins, 1946).

Millar, Stuart. On the Greenwich Millennium Clock in the *Guardian* (5 April 1997).

Opie, Iona & Peter. "Children's Calendar," in *The Lore and Language of Schoolchildren* (Clarendon Press, Oxford, 1959).

Ortiz, Alfonso. *The Tewa World: Space, Time, Being and Becoming in a Pueblo Society* (University of Chicago Press, 1969).

Palmer, Geoffrey & Noel Lloyd. *A Year of Festivals: A Guide to British Calendar Customs* (Warne & Co., 1972).

Russ, Jennifer M. *German Festivals and Customs* (Oswald Wolff, 1982).

Ryle, John. On the Summer Institute of Linguistics, in the *Guardian* (1997).

Staunton, Michael. On the year 1000, in the *Guardian* (31 December 1998).

Thomas, Hugh. *The Spanish Civil War* (Harper and Row, 1961).

Thomas, Keith. *Religion and the Decline of Magic: Studies in Popular Beliefs in Sixteenth and Seventeenth Century England* (Penguin, 1978).

Thompson, Damian. *The End of Time: Faith and Fear in the Shadow of the Millennium* (Sinclair-Stevenson, 1996).

Various authors. *The Land Is Ours Newsletter.*

Yamamoto, Yoshiko. *The Namahage: A Festival in the Northeast of Japan* (The Institute for the Study of Human Issues, 1978).

Zerubavel, Eviatar. *Hidden Rhythms: Schedules and Calendars in Social Life* (University of Chicago Press, 1981).

5. WREAKING GOOD HAVOC—A TIME OF WOMEN

Berger, John. *Ways of Seeing* (BBC & Penguin Books, 1972).

Briscoe, Joanna. *Skin* (Phoenix House, 1997).

The Cambridge Private Hospital (Leaflet) *Rhytidectomy: Surgical Facelift.*

Carroll, Helen. On child beauty queens, in the *Daily Mail* (5 June 1997).

Eliade, Mircea. *Patterns in Comparative Religion* (Sheed & Ward, 1958).

Ellis, Havelock. Quoted in *The Wise Wound,* Penelope Shuttle and Peter Redgrove (Victor Gollancz, 1978).

Falcon, Lidia. *Cartas a Una Idiota Española,* with cartoons by Nuria Pompeia (Dirosa, 1974).

Forman, Frieda Johles & Caoran Sowton. *Taking Our Time: Feminist Perspectives on Temporality* (Pergamon Press, 1989).

Fox, Meg. "Unreliable Allies: Subjective and Objective Time in Childbirth," in *Taking Our Time* (eds.) Forman and Sowton.

Greer, Germaine. *The Change* (Hamish Hamilton, 1991).

Heraclitus. *Fragments* (University Press of America, 1995).

Jung, Carl. Quoted in *About Time* (ed.) Christopher Rawlence.

Pitt-Rivers, Julian. *The People of the Sierra* (University of Chicago Press, 1966).

Roddick, Anita. *Body and Soul* (Ebury Press, 1991).

Shaw, Nancy. *Forced Labor: Maternity Care in the United States* (Pergamon Press, 1976).

Shelton, Beth Anne. *Women, Men and Time: Gender Differences in Paid Work, Housework and Leisure* (Greenwood Press, 1992).

Shiva, Vandana. *Staying Alive: Women, Ecology and Development* (Zed Books, 1988).

Shuttle, Penelope & Peter Redgrove. *The Wise Wound: Menstruation and Everywoman* (Victor Gollancz, 1978).

Taggart, James M. *Enchanted Maidens: Gender Relations in Spanish Folktales of Courtship and Marriage* (Princeton University Press, 1990).

Weiner, Annette B. *Women of Value, Men of Renown. New Perspectives in Trobriand Exchange* (University of Texas Press, 1976).

Valkeapää, Nils-Aslak. *The Sun My Father* (trans.) Ralph Salisbury, Lars Nordström, and Harald Gaski (University of Washington Press 1997).

Augustine, Saint. *The City of God*.

(Question: "What is time?" Answer: "If no one asks me, I know. If I am asked, I do not know.") *Confessions XI*, 14.

Bacon, Francis. *The Masculine Birth of Time* (1602).

Bohm, David. *Causality and Chance in Modern Physics* (Routledge & Kegan Paul, 1984).

Capra, Fritjof. *The Tao of Physics* (Wildwood House, 1975).

———. *The Web of Life: A New Synthesis of Mind and Matter* (Flamingo, 1996).

———. "Recent Research of Ilya Prigogine: A Summary" (private correspondence).

Coveney, Peter & Roger Highfield. *The Arrow of Time;* introduction by Ilya Prigogine (Flamingo, 1991).

Franchetti, Mark. On MiG fighter plane calendars, in the *Sunday Times* (1 June 1997).

Greer, Germaine. *The Female Eunuch* (MacGibbon & Kee, 1970).

Hawking Stephen. *A Brief History of Time* (Bantam, 1988).

Hieatt, Kent. *Short Time's Endless Monument: The Symbolism of the Numbers in Edmund Spenser's "Epithalamion"* (Columbia University Press, 1960).

Hundertwasser, Friedensreich. *The Beauty of Fractals* (Peitgen, H. O. & Richter, P. H. Springer-Verlag, 1986).

Keller, Evelyn F. *Reflections on Gender and Science* (Yale University Press, 1985).

Kristeva, Julia. "Women's Time" in *The Kristeva Reader* (ed.) Toril Moi (Blackwell, 1986).

Krudy, E., B. Bacon, & R. Turner. *Time: A Bibliography* (Information Retrieval Ltd., 1976).

Lovelock, James. *The Ages of Gaia: A Biography of Our Living Earth* (Oxford University Press, 1995).

Malleus Maleficarum (1486).

Patrides, C. A. (ed.) *Aspects of Time* (Manchester University Press, 1976).

Pope, Alexander. *Collected Poems* (J. M. Dent & Sons, 1963).

Prigogine, Ilya & Isabelle Stengers. *Order Out of Chaos: Man's New Dialogue with Nature* (Heinemann, 1984).

Spenser, Edmund. "Epithalamion" (1595).

Steiner, George. *After Babel* (Oxford, 1975).

Stewart, Ian. *Does God Play Dice? The Mathematics of Chaos* (Blackwell, 1989).

Virgil. *Æneid,* Book 4

Wakeford, Tom & Martin Walter. (eds.) *Science for the Earth: Can Science Make the World a Better Place?* (John Wiley, 1995).

Zohar, Danah. *Through the Time Barrier: A Study of Precognition and Modern Physics* (Heinemann, 1982).

7. THE POWER AND THE GLORY

Achebe, Chinua. *Arrow of God* (Heinemann, 1964).

Ballard, J. G. *The Four-Dimensional Nightmare* (Victor Gollancz, 1963).

Bentham, Jeremy. *Panopticon* (1791).

Buckley, Jerome Hamilton. *The Triumph of Time: A Study of the Victorian Concepts of Time, History, Progress and Decadence* (Belknap Press, Harvard University Press, 1967).

Cameron, Deborah. *The Feminist Critique of Language: A Reader* (Routledge, 1990).

Canetti, Elias. *Crowds and Power* (Victor Gollancz, 1962).

Clayton, Rev. J. ("of Brazen Nose College, Oxford"). *Friendly Advice to the Poor* (Manchester, 1755).

Conrad, Joseph. *The Secret Agent* (1906).

Coventry Telegraph, On the amount of time children spend with parents (15 July 1997).

Defoe, Daniel. *Robinson Crusoe* (1719).

Dickens, Charles. *Hard Times* (1854).

Duerr, Hans Peter. *Dreamtime: Concerning the Boundary Between Wilderness and Civilization* (Blackwell, 1985).

Duncan, David Ewing. *The Calendar* (Fourth Estate, 1998).

Duveau, Georges. *La Vie Ouvrière en France* (Paris, 1946).

Elias, Norbert. *Time: An Essay* (Blackwell, 1987).

Ende, Michael. *Momo* (K. Thienemanns Verlag, 1973).

Firth, Raymond. *The Work of the Gods in Tikopia* (Athlone Press, 1967).

Fabian, Johannes. *Time and the Other: How Anthropology Makes Its Object* (Columbia University Press, 1983).

Foster, John. *An Essay on the Evils of Popular Ignorance* (Holdsworth, London, 1820).

Gell, Alfred. *The Anthropology of Time: Cultural Constructions of Temporal Maps and Images* (Berg, 1992).

Greenhouse, Carol J. *A Moment's Notice: Time Politics Across Cultures* (Cornell University Press, 1996).

The Guardian. On time spent in cars (23 July 1997).

Hall, Edward, T. *The Silent Language* (Doubleday, 1959).

Harmon, William. *Time in Ezra Pound's Work* (University of North Carolina Press, 1977).

Hobsbawm, Eric & Terence Ranger. *The Invention of Tradition* (Past & Present Publications, Cambridge University Press, 1983).

Holloway, John & Eloina Pelaez. *Zapatista! Reinventing Revolution in Mexico* (Pluto Press, 1998).

Kipling, Rudyard. "If" (1910).

Kincaid, Jamaica. *A Small Place* (Virago, 1988).

Lewis, Nigel. *The Book of Babel* (Penguin, 1995).

Maclean, Charles. *Island on the Edge of the World: The Story of St Kilda* (Canongate, Edinburgh, 1977).

Marcuse, Herbert. *Eros and Civilization* (Routledge and Kegan Paul, 1956).

Marx, Karl. *Das Kapital* (Hamburg, 1867).

Mumford, Lewis. *The Human Prospect* (Secker & Warburg, 1956).

———. *Technics and Civilization* (Routledge & Sons, 1934).

Narby, Jeremy. *Visions of Land: The Ashaninca and Resource Development in the Pichis Valley in the Peruvian Central Jungle.* (Ph.D. dissertation, Stanford University. Ann Arbor: University Microfilms, 1989).

Nowotny, H. *Time: The Modern and Postmodern Experience* (Polity Press, 1994).

"Running Out of Time," television documentary, 1994. KCTS (Seattle) and Oregon Public Broadcasting.

Spender, Dale. *Man Made Language* (Routledge & Kegan Paul, 1980).

Steiner, George. *Antigones* (Clarendon Press, Oxford, 1984).

Time (magazine), August 1997.

Trolls, Various. *Manifesto of the Independent Free State of Trollheim.* Broadcast on Tree FM pirate radio (1997).

Wordsworth, William. *The Prelude* (1805).

Wesley, John. *The Duty and Advantage of Early Rising* (London, 1786).

Zeldin, Theodore. *An Intimate History of Humanity* (Sinclair Stevenson, 1994).

Zimmerman, Don & Candace West. in *Language and Sex. Difference and Dominance,* (eds.) B. Thorne & N. Henley (Newbury House, Rowley, Mass. 1975).

8. LIFE'S TOO SHOR

Anon. *The Cloud of Unknowing* (c. 1370).

Baudrillard, Jean. *Selected Writings* (Polity Press, 1988).

Baxter, R. *A Christian Directory* (R. White, London, 1673).

BBC Radio 3, 21 May 2001, on jet-lag shrinking the brain.

Borges, Jorge Luis. *Labyrinths* (James Laughlin, 1962).

Black, Bob. *The Abolition of Work* (Pirate Press, 1990) (c/o *Neither Work nor Leisure,* PO Box HP94 Leeds, LS6 1YJ).

Coser, Lewis. *Greedy Institutions* (Free Press, 1974).

Cuautémoc, Guaicaipuro. "The Real Foreign Debt" in *Resurgence* No. 184 and in *Revista Renacer Indianista* No. 7.

Ellis, Gareth. On electronic timesheets in *The Idler* (August/September 1998).

Franklin, Benjamin. *Advice to a Young Tradesman* (1748).

Future Foundation. Report on twenty-four hour society (November 1997).

Gabriel, Yiannis & Tim Lang. *The Unmanageable Consumer* (Sage Publications, 1995).

Ghazi, Polly & Judy Jones. *Getting a Life: The Downshifter's Guide to Happier, Simpler Living* (Hodder & Stoughton, 1997).

Gorz, André. *Farewell to the Working Class: An Essay on Post-Industrial Socialism* (Editions Galilee, 1980).

———. *Critique of Economic Reason* (Verso, 1989).

The Guardian, 13 September 2001, on the World Trade Center.

Hager, L. Michael. "The Nonstop City," in *The Futurist* (May/June 1997).

Handy, Charles. *The Empty Raincoat* (Arrow Books, 1994).

Harrold, C. F. *John Henry Newman* (Longman, 1945).

Hood, Jane C. (ed.). *Men, Work and Family* (Sage Publications, 1993).

International Herald Tribune, 29 March 2001, on the National Sleep Foundation.

Lebow, Victor. In the *New York Journal of Retailing* (1955).

LeGoff, Jacques. *Time, Work and Culture in the Middle Ages* (Chicago University Press, 1980).

Lippman, John. "Television a Channel for World Change," in the *Detroit News* (1 November, 1992).

Mattox, William R., Jr. "America's Family Time Famine," in *Children Today* (Nov./Dec. 1990).

Melbin, Murray. *Night as Frontier: Colonizing the World After Dark* (Free Press, 1987).

Mirror, 12 September 2001, on the World Trade Center.

The New York Times, 13 September 2001, on the World Trade Center.

New Road Map Foundation, *All-Consuming Passion: Waking Up from the American Dream* (Seattle).

Radio 4, figures on American night work (8 April 1999).

Robinson, John P. *I Love My TV: The Demographics of Time Use* (Ithaca, NY: American Demographics, 1994).

Ruskin, John. *Time and Tide* (Smith, Elder & Co, 1867).

Saint-Exupéry, Antoine de. *The Little Prince* (Heinemann, 1944).

Schor, Juliet. *The Overworked American* (Basic Books, 1992).

Shakespeare, William. *Richard II.*

——. *Troilus and Cressida.*

Spengler, Oswald. *The Decline of the West* (Allen & Unwin, 1926).

Spry, Irene M. In *Consuming Canada: Readings in Environmental History* (eds.) Gaffield and Gaffield (Copp Clark, Toronto, 1995).

Sterne, Laurence *The Life and Opinions of Tristram Shandy* (1759).

Taylor, Frederick Winslow. *The Principles of Scientific Management* (Harper & Bros, 1911).

Third World First literature on the Third World Debt (1998).

Thomas, Richard. On the lost weekend, in the *Observer* (11 October 1998).

Waskow, Rabbi Arthur. In *The Nation,* 1 January 2001.

Younge, Gary. On the suicide of Christopher Bryant, in the *Guardian* (14 March 1997).

9. PROGRESS IS A FOUR-LETTER WORD

The Balaton Group. On the US Patent Office (Summer 1997).

Darwin, Charles. *On the Origin of Species* (John Murray, 1859).

——. *The Descent of Man* (John Murray, 1871).

Durkin, Martin (producer). *Against Nature* (Production company: RDF. Broadcast Channel 4, 1997).

Ereira, Alan. *The Heart of the World* (Jonathan Cape, 1990).

Falla, Jonathan. *True Love and Bartholomew* (Cambridge University Press, 1991).

Fox, Warwick. "Human Empire" in *Resurgence,* No. 184.

George, Henry. *Progress and Poverty* (Appleton & Co., 1880).

Gilpin, William. *Mission of the North American People: Geographical, Social and Political* (Philadelphia, 1873).

Gyatso, Palden. *Fire Under the Snow: Testimony of a Tibetan Prisoner,* introduction by Tsering Shakya (Harvill, 1997).

Kumar, Satish. *No Destination: An Autobiography* (Resurgence, 1992).

Krenak, Ailton. Quoted in Liz Hosken, *A Tribute to the Forest People* (Gaia Foundation, 1990).

Leakey, Richard & Roger Lewin. *The Sixth Extinction: Biodiversity and its Survival* (Weidenfeld & Nicolson, 1996).

Lewington, Anna, and Edward Parker. *Ancient Trees: Trees That Live for a Thousand Years* (Collins and Brown, 1999).

Lindqvist, Sven. *Exterminate All the Brutes* (Granta, 1997).

Lopez, Barry. *Arctic Dreams: Imagination and Desire in a Northern Landscape* (Scribner's, 1986).

Moody, Roger (ed.). *The Indigenous Voice* (Zed & International Work Group for Indigenous Affairs, 1988).

Prime, Ranchor. *Hinduism and Ecology* (Cassell, 1992).

Radford, Tim. On the Biosphere experiment, in the *Guardian* (6 March 1997).

Roberts, J. M. *The Pelican History of the World* (Pelican, 1988).

Sahlins, Marshall. *The Use and Abuse of Biology* (Tavistock Publications, 1977).

Saro-Wiwa, Ken. *Genocide in Nigeria: The Ogoni Tragedy* (Saros International Publishers, 1992).

Sasubrin, Vladimir. Quoted in *Spiegel* (1 April 1990).

Schumacher, E. F. *Small Is Beautiful: A Study of Economics as if People Mattered* (Blond & Briggs, 1973).

Schwartz, Hillel. Figures on species extinction from *Century's End: An Orientation Manual Toward the Year 2000* (Doubleday, 1990).

Shiva, Vandana. *In The Future of Progress: Reflections on Environment and Development* (ed.) Edward Goldsmith et al. (International Society for Ecology and Culture, 1992).

Tobert, Natalie. "Rainforest Shamans" in *The Ecologist* (July/August 1998).

Vidal, John. On the U'wa people of Colombia in the *Guardian* (20 September 1997).

———. On fog in Southeast Asia in the *Guardian* (27 September 1997).

Wong, Datuk Amar James K. M. *Buy a Little Time.*

10. A TEFLON TOMORROW

Abram, David. *The Spell of the Sensuous* (Vintage, 1997).

Berman, Marshall. *All That Is Solid Melts Into the Air: The Experience of Modernity* (Verso, 1983).

Burke, Edmund. *Reflections on the Revolution in France* (1790).

Brown, Paul. On hormone disrupting chemicals, in the *Guardian* (28 January 1998).

Burgess, Anthony. *Honey for the Bears* (Heinemann, 1963).

Butler, Samuel. *Erewhon* (Trübner & Co., 1872).

Castells, Manuel. *The Information Age: Economy, Society and Culture,* 3 Vols. (Blackwell, 1996–98).

Colborn, T., J. P. Myers, & D. Dumanoski, *Our Stolen Future: How Manmade Chemicals Are Threatening Our Fertility, Intelligence and Survival* (Little, Brown, 1996).

Council for Posterity in London. *Declaration of the Rights of Posterity* (London, 1990).

Crystal, David. *English as a Global Language* (Cambridge University Press, 1997).

The Daily Record, 8 June 1998, on Dounreay.

Eliade, Mircea. *Shamanism* (Routledge & Kegan Paul, 1964).

Evans, Paul. On space mirror technology, in the *Guardian* (10 February 1999).

Forum 2000 *Conference Report 1997* (Prague, 1997).

Gagnon, Bruce. (Co-ordinator of the Florida Coalition for Peace and Justice). Interview concerning Cassini in *Vista* (July 1997).

Goethe, Johann Wolfgang von. *Faust.*

Glendinning, Chellis. "Notes Toward a Neo-Luddite Manifesto," *The Utne Reader* (March 1990).

Kermode, Frank. *The Sense of an Ending* (Oxford University Press, 1966).

Krauss, Michael. *Language,* Vol. 68, No. 1.

McLuhan, Marshall. *The Gutenberg Galaxy* (Routledge & Kegan Paul, 1962).

Miller, Norman. On terraforming, in the *Independent on Sunday* (12, October 1997).

Radford, Tim, On the colonization of space, in the *Guardian* (18 October 1997).

Rich, Adrienne. *Leaflets: Poems 1965–1968* (Chatto & Windus, 1972).

Roddick, Anita. On the Cassini space probe, in the *Guardian* (8 October 1997).

Sale, Kirkpatrick. *Rebels Against the Future: The Luddites and Their War on the Industrial Revolution: Lessons for the Computer Age* (Addison-Wesley, 1995).

Shakespeare, William. *King Henry IV, Part 1.*

———. *The Merchant of Venice.*

Shell (International Petroleum Company). *Global Scenarios 1992–2020* (Shell, 1992). Not to be confused with the far more entertaining publication on Shell and the Ogoni by "Sam and Daniel" *They Seem So Friendly™; Shell on Safari* (Play Publications, PO Box 1102,1895 Commercial Drive, Vancouver, BC, Canada).

Shelley, Mary. *Frankenstein or the Modern Prometheus* (1818).

Slouka, Mark. *War of the Worlds: Cyberspace and the High-Tech Assault on Reality* (Abacus, 1996).

Steiner, George. *George Steiner: A Reader* (Penguin, 1984).

Toffler, Alvin. *Future Shock* (Bodley Head, 1970).

Wilson, Edward O. *Biophilia* (Harvard University Press, 1984).

World Future Society. *Future Vision* (Bethesda, Maryland, USA, 1996).

www.theonion.com.

www.terralingua.org.

11. NATOURE™

Arlidge, John. On Joseph Rotblat and cloning, in the *Guardian* (26 February 1997).

Baudelaire, Charles. *The Complete Poems* (Limouse Museum, 1992).

Baumann, M., J. Bell, F. Koechlin, & M. Pimbert (eds.). *The Life Industry: Biodiversity. People and Profits* (Intermediate Technology Publications, 1996).

Easterbrook, Gregg. *A Moment on the Earth: The Coming Age of Environmental Optimism* (Penguin, 1995).

Fawcett, Henry. *Manual of Political Economy* (Cambridge, 1863).

Franklin, A. *La Mésure de temps* (Paris, 1888).

The Genetics Forum: details on Terminator Technology.

Jones, Steve. *The Language of the Genes* (Flamingo, 1993).

Kitcher, Philip. *The Lives to Come: The Genetic Revolution and Human Possibilities* (Allen Lane, 1996).

King, David. In *GenEthics News*.

Marvell, Andrew. "To His Coy Mistress" (1650–2).

McKie, Robin. On the cloning of Dolly, in the *Observer* (23 February 1997).

Pollack, Robert. *Signs of Life: The Language and Meanings of DNA* (Viking, 1994).

Rifkin, Jeremy. *The Biotech Century: Harnessing the Gene and Remaking the World* (Victor Gollancz, 1998).

Shiva, Vandana. *Monocultures of the Mind: Perspectives on Biodiversity and Biotechnology* (Zed Books, 1993).

———. In *The Life Industry* (eds.) Baumann, Bell, Koechlin & Pimbert.

Silver, Lee M. *Remaking Eden: Cloning and Beyond in a Brave New World* (Weidenfeld & Nicolson, 1998).

Standing Bear, Chief Luther. *Land of the Spotted Eagle* (Houghton Mifflin, 1933).

Steinbrecher, Dr. Ricarda A. On genetic engineering, in *The Ecologist* (November/December 1996).

Vidal, John. On the world's first patented man, in the *Guardian* (12 November 1994).

Virgil. *Eclogues*

———. *Georgics*.

von Weizsäcker, Christine. "Biodiversity Newspeak," in *The Life Industry* (eds.) Baumann, Bell, Koechlin & Pimbert.

Wilson, Edward O. *The Diversity of Life* (Allen Lane, 1992).

Albery, N., G. Elliot, & J. Elliot (eds.). *The Natural Death Handbook* (Virgin, 1993).

Ariès, Philippe. *Western Attitudes Toward Death From the Middle Ages to the Present* (Johns Hopkins University Press, 1974).

Barley, Nigel. *Dancing on the Grave: Encounters with Death* (John Murray, 1995).

Bede. *Ecclesiastical History of the English People* (731).

Benjamin, Marina. *Living at the End of the World* (Picador, 1998).

Boseley, Sarah & Ed Vulliamy. On Trans Time, in the *Guardian* (1 March 1997).

Eire, Carlos M.N. *From Madrid to Purgatory* (Cambridge University Press, 1995).

Gill, Sue & John Fox. *The Dead Good Funerals Book* (Engineers of the Imagination, Welfare State International, Ulverston, Cumbria, 1997).

Gorer, Geoffrey. "The Pornography of Death" (1955). Quoted in Philippe Ariès' *Western Attitudes Toward Death*.

Hegel. Quoted in George Steiner's *Antigones*.

Ihimaera, Witi. *Tangi* (Heinemann, 1973).

Illich, Ivan. *Tools for Conviviality* (Harper and Row, 1973).

Margolis, Jonathan. On the Immortal Genes Co., in the *Evening Standard* (16 April 1997).

Marquis, Don. *archy and mehitabel* (Heinemann, 1927).

Muir, John. *A Thousand-mile Walk to the Gulf* (Houghton Mifflin Company, 1916).

Nuland, Sherwin B. *How We Die* (Chatto & Windus, 1994).

Rose, Kenneth Jon. *The Body in Time* (Wiley, 1988).

Solomon, Deborah. "Mourning Becomes Electric," *Detroit Free Press,* 27 October 1996.

Whitman, Walt. *Leaves of Grass* (Washington, 1872).

Ahuja, Anjana. On experimental psychology in *The Times* (30 March 1998).

Angelou, Maya. *Conversations with Maya Angelou* (Virago, 1989).

Baudelaire, Charles. "On Getting Drunk," quoted in *The Vintage Book of Dissent* (eds.) Michael Rosen and David Widgery (Vintage, 1996).

Becker, J. "Hindu-Buddhist Time in Javanese Gamelan Music," in *The Study of Time IV* (eds.) J. T. Fraser, N. Lawrence & D. Park (Springer, 1981).

Clark, Kenneth. *Civilisation* (BBC, 1969).

Donne, John. "The Sun Rising" from *Songs and Sonnets* (c. 1595–1605).

Franz, Marie-Louise von. *Time: Rhythm and Repose* (Thames & Hudson, 1978).

Grandison, Bishop of Exeter. "Mandate Against Gigglings," quoted in *The Vintage Book of Dissent* (eds.) Michael Rosen and David Widgery (Vintage, 1996).

Gwilym, Dafydd ap. *Poems* (ed.) Richard Morgan Loomis (Center for Medieval & Early Renaissance Studies, S.U.N.Y. 1982).

Hui-neng. Quoted in Capra's *The Tao of Physics*.

Huizinga, Johan. *Homo Ludens: A Study of the Play Element in Culture* (Routledge & Kegan Paul, 1949).

Jung, Carl. *Man and His Symbols* (Aldus Books, 1964).

Kakar, Sudhir. *The Inner World: A Psycho-analytic Study of Childhood and Society in India* (Oxford University Press, 1978).

Lao-Tzù. *Tao Te Ching*.

Leach, Edmund. "Time and False Noses" in *Rethinking Anthropology* (Athlone Press, 1961).

Lindqvist, Sven. *The Skull Measurer's Mistake: and Other Portraits of Men and Women Who Spoke Out Against Racism* (The New Press, 1997).

Marshall, Robert. Memorandum to Harold Ickes, Feb 27, 1934, Record Group 79 (National Park Service).

McPhee, John. *Basin and Range* (Farrar Straus Giroux, 1980).

Muir, John. *Travels in Alaska* (Houghton Mifflin, 1915).

National Archives, Washington, D.C.

Paul, Saint. *I Corinthians* 7:9.

Russell, Bertrand. *In Praise of Idleness* (Allen & Unwin, 1935).

Service, Robert. *Collected Poems of Robert Service* (Dodd, Mead & Co., 1964).

Suzuki, Daisetz Teitaro. *Zen and Japanese Culture* (Routledge & Kegan Paul, 1959).

Van der Post, Laurens. *Jung and the Story of Our Time* (Hogarth Press, 1976).

The author and publisher of this work are grateful to the Center for Medieval and Early Renaissance Studies for permission to quote from *Poems* by Dafydd ap Gwilym, edited by Richard Morgan Loomis.

EXHIBITIONS

The Hayward Gallery, *The Art of Ancient Mexico* (1992)

Museum of Mankind, *Mexican Day of the Dead* (1991)

The Alaska State Museum, Juneau, Alaska (1997)

Melanie Manchot at The Octagon Galleries, Bath (1997)

Robert Maclaurin, New Paintings of Australia, Berkeley Square Gallery (1997)

The world's largest maze, Frilford near Abingdon in Oxfordshire, designed by Adrian Fisher (1997).

SOUNDTRACKS

"Designer Kidz" from *Seize the Day* by Theo Simon and Shannon Smy. Seize the Day, PO Box 23, 5 High Street, Glastonbury, Somerset BA6 9DP

The River League
*the wilderness rafting company which campaigns for the protection of the Taku
and other endangered rivers.*
Ste 201–1112 Broughton St., Vancouver,
British Columbia, Canada. V6G 2A8
e-mail: *iankean@riverleague.ca*
www.riverleague.ca
Tel: 604 687 3417
Fax: 604 687 3413

The Taku Wilderness Association
Box 142, Atlin, BC VOW 1AO
Tel/Fax: 250 651 0047
e-mail: *twa@ibm.net*
www.taku.org

Two organizations involved in the campaign for the U'wa are:
I.W.G.I.A.
Fiolstraede 10, Copenhagen,
DK 1171, Denmark
Tel: (01145) 33 124724
Fax: (01145) 33 147749

O.N.I.C.
(the National Organization for Indigenous Peoples in Colombia)
Calle 13
No. 4/38, Bogotá,
AA 32395, Colombia
Tel: (011571) 2428017
Or: (011571) 2848196

The Land Is Ours
Britain's land rights campaign
Box E, 111 Magdalen Road, Oxford
OX4 1RQ
Tel: 1865 722 016

Reclaim the Streets
PO Box 9656, London N4 4JY
rts@gn.apc.org

Totnes Genetics Group,
local and national campaigning against genetic engineering
TOGG
Applebarn, Week, Dartington,
Near Totnes, Devon, TQ9 6JP
Tel: 1803 840098
www.togg.freeserve.co.uk/core.htm
http.//visitweb.com/totnes

The Earth Centre
exhibiting and campaigning for sustainability
Denaby Main
Doncaster
DN12 4EA
Tel: 1709-512000
e-mail: *info@earthcentre.org.uk*
www.earthcentre.org.uk

Survival International
campaigning for the rights of indigenous people
11–15 Emerald Street
London WC1 3QL
Tel: 171 242 1441

Common Ground
campaigning for cultural landscapes and local distinctiveness
PO Box 25309
London NW5 1ZA
Tel: 171 267 2144

The Soil Association
campaigning for organic food, farming and sustainable forestry
Bristol House
40–56 Victoria Street
Bristol BS1 6BY
Tel: 117 929 0661

INTERVIEWS

The following are only a handful of those I talked to about the book, but a very important handful.

Amilton Lopez (Ava Pykavera) and Rossolino Ortiz from the Guaraní-Kaiowa tribes of Mato Grosso do Sul, Brazil

Daniel Zapata, spokesperson for the Navajo and Hopi peoples

George Monbiot, environmental activist, journalist and author

Richard Gott, journalist, author and expert on Jesuit history and imperialism

Alan Tormaid Campbell, anthropologist and author

Edward Goldsmith, Editor, *The Ecologist*

Trainer Tate, Maori freedom fighter

Jonathan Raban, novelist and editor of *The Oxford Book of the Sea*

John Whitelegg, Professor of the Built Environment, Liverpool John Moores University

Mayer Hillman, Senior Fellow Emeritus, Policy Studies Institute

Steve Bell, cartoonist

Martin Rowson, cartoonist

King Arthur Uther Pendragon, himself

Dr. Oliver Rackham, Woodland Ecologist at Corpus Christi College, Cambridge

Martin Drury, Director General of the National Trust

Jocelyn Stevens, Chair of English Heritage

Dr. Barbara Bender, Anthropology Department, University College, London

Roger Rainbow and Ged Davis, Shell International Petroleum Company

Colin Evans, past life regression hypnotist

Mark Marchant, racing driver at Brands Hatch

Jan Tritten, editor of *Midwifery Today*

Lisa Cox, National Women's Health Network, Washington

Mark Edwards, author and photographer, *Still Pictures*

Victor Anderson, author of *Alternative Economic Indicators*

Professor Gwynfor Jones, History Department, Cardiff University

INDEX

ACKNOWLEDGMENTS

RESPECT

This is a public place for private salutes. I have to write this. (You don't, of course, have to read it.) This book took years to write and these are the people without whom . . .

Respect to my grandmothers, on my father's side for her generosity and largesse, on my mother's for her love of language and books. Respect to my father for your idiosyncratic thinking, to my mother for your beautiful values; pleasure, honesty and good compost. Respect to my brothers for years of making me giggle, and for your love.

Respect to Alison for your quarter-century of friendship. Respect to Rob for fighting my corner like a brother. To Jan for your huge enthusiasms and *joie de vivre*. To George, for the breadth of your interests, your brilliant polemics, for your thoughtful textual suggestions and for your friendship. To Giuliana for being so young and so old, to Julia for your warmth, humanity and tea. Respect to Clare for saying what I thought *I* was about to say, for your laugh-factor, to Luci for such long belief in this book. And to Vic my thanks for the why which you know, for such utter kindness, for your open-hearted generosity of time.

To Margaret and Alex, my thanks for the island-off-the-island refuge and for your welcome there. To Adam, Simon and Fiona for being so intelligent, so funny and so damn nice. To Michael, Sophie and Alex and to Bugs and Nancy and to Adrian for piling on the positives. Thanks to Anita, Gordon and Sam for the Taku, to Ian for visionary and poetic protection of wildernesses. Thanks to Hannah for the wettest canoeing. Thanks to a whole heap of protesters for your style, your politics and your wit. Thanks to Satish for bringing Indian sagacity to England. Thanks to Fritjof for your generous encouragement and, now as ever, such respect for writing *The Tao of Physics* and *The Web of Life*. Respect to Sureyani, for your spirit.

And Mr Social Control. You are my infinity.

Thank you to Barbara Moulton, for such judicious advice and wise agenting. Thank you to Sara Carder for appreciative and sensitive editing.

Thanks to Liz and my American family for Americana and help.

To Jeremy. You threw me a lifeline, you believed in the ideas of this book, you shared its spirit, you were its ambassador. I can not thank you more.

And a dedication. This book was inspired by one man. One extraordinary one. He first suggested that I should write about time, and for a long time writing this book he was all my audience. It grew in conversations with him, he watched over its many stages. It would never have occurred to me to take on a subject so huge, but from the beginning his was the confidence, his the utter belief, his the boundless enthusiasm, his the magnificence, his the love and his the conjuring. He made the world bigger to me. And then he filled it with color. It began with him then, it ends with him now. With my love, this book could only be dedicated to: John Vidal.

DATE			